BASIC CONCEPTS IN CHEMISTRY

Functional Group Chemistry

J.R. HANSON
University of Sussex

WILEY-
INTERSCIENCE

RS•C
ROYAL SOCIETY OF CHEMISTRY

Cover images © Murray Robertson/visual elements 1998–99, taken from the 109 Visual Elements Periodic Table

Library of Congress Cataloging-in-Publication Data:
Library of Congress Cataloging-in-Publication Data is available.
ISBN: 0-471-22480-4

Typeset in Great Britain by Wyvern 21, Bristol
Printed and bound by Polestar Wheatons Ltd, Exeter

10 9 8 7 6 5 4 3 2 1

Preface

The aim of this book is to provide an introduction to the characteristic properties of functional groups. It is written for first-year undergraduates.

The book is in four chapters. The first is devoted to a general consideration of the bonding in functional groups, the classes of reagent and the types of reaction. Functional groups may be divided into several broad classes. The first of these are those functional groups in which the reactions, mainly substitution and elimination, are those of the σ-bond. The chemistry of these functional groups forms the second chapter. A second class of functional groups is those in which a π-bond is a characteristic feature. The initial step in many of their reactions is an addition. These functional groups are described in the third chapter. The electrons within a π-bond may be symmetrically distributed as in an alkene, or unsymmetrically distributed as in the carbonyl group. The aromatic ring plays a major part in functional group chemistry. The final chapter is devoted to a description of the interaction between functional groups and the aromatic ring. Heteroaromatic compounds are considered in terms of the perturbation of the π-system brought about by the insertion of the heteroatom.

A balance has to be drawn in the use of systematic and trivial names. The IUPAC rules recognize that the use of trivial names for many simple compounds is often preferred. Students will meet these names in reading the current literature and on the bottles in the laboratory. However, because the systematic nomenclature forms the basis of documenting more complex structures, students need to be familiar with the use of systematic nomenclature by applying it to simple molecules. Both names will be given for compounds at appropriate stages in the text. An Appendix of Common and Systematic Names and Further Reading list can be found on the Wiley website at www.wiley.com/go/wiley-rsc.

I am indebted to Martyn Berry and Professor Sir John Cornforth AC FRS for their many valuable comments on the draft manuscript and particularly to Professor Alwyn Davies FRS for his substantial help and encouragement throughout the preparation of the manuscript and diagrams.

J. R. Hanson
Sussex

BASIC CONCEPTS IN CHEMISTRY

This series of books consists of short, single-topic or modular texts, concentrating on the fundamental areas of chemistry taught in undergraduate science courses. Each book provides a concise account of the basic principles underlying a given subject, embodying an independent-learning philosophy and including worked examples. The one topic, one book approach ensures that the series is adaptable to chemistry courses across a variety of institutions.

TITLES IN THE SERIES

Stereochemistry *D G Morris*
Reactions and Characterization of Solids *S E Dann*
Main Group Chemistry *W Henderson*
d- and f-Block Chemistry *C J Jones*
Structure and Bonding *J Barrett*
Functional Group Chemistry *J R Hanson*

Further information about this series is available at www.wiley.com/go/wiley-rsc

Contents

1 General Principles

Aims

The aims of the first chapter of this book are to provide the foundations for functional group chemistry. By the end of this chapter you should be able to understand:

- The relationship between bonding and structure of organic compounds
- The oxidative and substitutive relationship between functional groups
- The relationship between electronegativity differences and the reactivity of functional groups
- The reactivity of nucleophilic, electrophilic and radical reagents
- The role of acids and bases in the catalysis of organic reactions
- The influence of electronic and steric factors on reactivity
- The kinetic and thermodynamic control of reaction products

1.1 The Structure of Functional Groups

A functional group is a chemically reactive group of atoms within a molecule. Each functional group has its characteristic reactivity, which may be modified by its position within the molecule or by the presence of other neighbouring functional groups.

1.1.1 Hybridization and Bonding

An isolated carbon atom possesses two electrons in its 1s orbital, two electrons in its 2s orbital and two electrons in its 2p orbitals. The types of bonding found in carbon compounds arise from various hybrids of the 2s and 2p orbitals. Combination of one 2s and three 2p orbitals

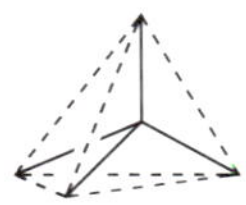

1.1

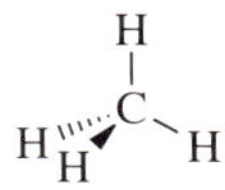

1.2

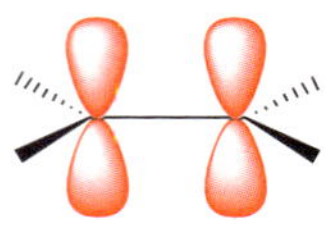

1.3

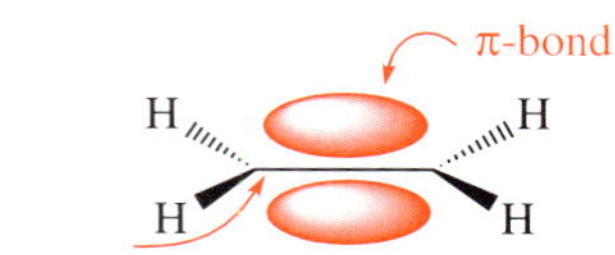

1.4

1.5

H−C≡C−H

1.6

leads to four equivalent **sp^3 hybridized** orbitals directed towards the apices of a tetrahedron (see **1.1**). Each of these orbitals can be occupied by one of the four available electrons from the carbon atom. These can each pair with one electron from another atom to produce, for example, methane (**1.2**). The energy required for the reorganization of the orbitals is gained from the formation of the four covalent bonds.

Alternatively, the 2s and two 2p orbitals may be hybridized to give a planar sp^2 system accommodating three electrons from the carbon, one in each hybrid orbital. Three bonds may then be formed with other atoms (see **1.3**). The remaining electron, which is in a p orbital at 90° to the plane of the sp^2 system, may overlap with a comparable p orbital from a second atom to form a **π-bond**, leading to a double bond between the carbon and this atom as in ethene (**1.4**).

A further way of making four bonds from the carbon is to hybridize the 2s and one 2p orbital to give two sp hybrids in which the orbitals are at 180° to each other. The remaining two 2p orbitals are used to form two π-bonds at 90° each other (see **1.5**). In this case there is a triple bond between the carbon and another atom as, for example, in ethyne (acetylene, **1.6**).

These hybridizations have several consequences. Since an s orbital is closer to the nucleus than the corresponding p orbital, the increasing s character in the orbitals in changing from sp^3 to sp^2 and then sp leads to a decrease in bond length: sp^3 C–C, 0.154 nm; sp^2 C=C, 0.134 nm; sp C≡C, 0.120 nm. Secondly, the increasing s character means that the bonding electrons in the sp and sp^2 orbitals are held more tightly to the carbon than in an sp^3 bond. This is reflected in the increase in the polarity and acidity of a $C^-–H^+$ bond and in the ease of formation of organometallic compounds containing the $C^-–M^+$ bond. These follow the order $sp^3 < sp^2 < sp$. On the other hand, there is an increase in the difficulty of breaking a carbon–halogen bond in the ionic sense $C^+–X^-$, in changing from an alkyl to an alkenyl (vinyl) halide. Thirdly, whereas the maximum electron density of a σ-bond lies between the atoms forming the bond, that of a π-bond lies above and below the plane of the bonding atoms, *i.e.* a π-bond is more exposed for reaction.

A further consequence lies in the opportunity for one π-bond to interact with another suitably oriented π-bond to give a **conjugated** system. When a carbon–carbon double bond is separated from another carbon multiple bond by one carbon–carbon single bond so that overlap is possible between the π-bonds, they are said to be conjugated. Thus in butadiene (**1.7**) molecular orbitals may be written embracing all four atoms. These conjugated double bonds often behave as one functional group rather than as two isolated double bonds. Electronic effects are relayed through the conjugated system.

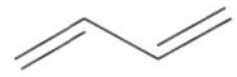

1.7

1.8

1.9

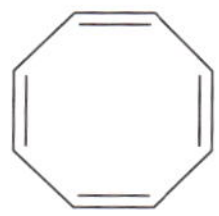

1.10

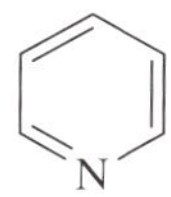

1.11

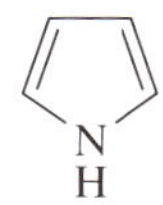

1.12

$H_2C{=}C{=}CH_2$

1.13

A cyclic conjugated system containing $(4n + 2)\pi$ electrons has an extra stability over that of a comparable number of isolated double bonds. This extra stabilization, known as **aromaticity**, leads to a characteristic pattern of reactivity which distinguishes the reactions of benzene (**1.8**) from, for example, the linear hexatriene (**1.9**) or cyclooctatetraene (**1.10**) ($4n$ electrons, $n = 2$). The aromatic sextet may arise not just from the overlap of three double bonds as in benzene (**1.8**) or pyridine (**1.11**) but also from the participation of the lone pair of electrons on a heteroatom. Thus pyrrole (**1.12**), with effectively six π-electrons, shows some aromatic character. In allene (**1.13**) the double bonds are at 90° to each other and conjugation does not occur.

1.1.2 Bonding and Structure

The tetrahedral arrangement of the bonds of an sp^3 hybridized carbon atom, the planar trigonal sp^2 arrangement and the linear sp system each have structural and geometrical consequences. The existence of free rotation about a single bond means that in a molecule such as ethane the methyl groups are free to take up a range of different **conformations** relative to each other. There are two extreme conformations, one in which the hydrogen atoms are staggered (see **1.14**) and the other in which they are eclipsed (see **1.15**). In the former the interactions between the hydrogen atoms of the methyl group are minimized, and the structure is of a lower energy than that in which the hydrogen atoms are eclipsed.

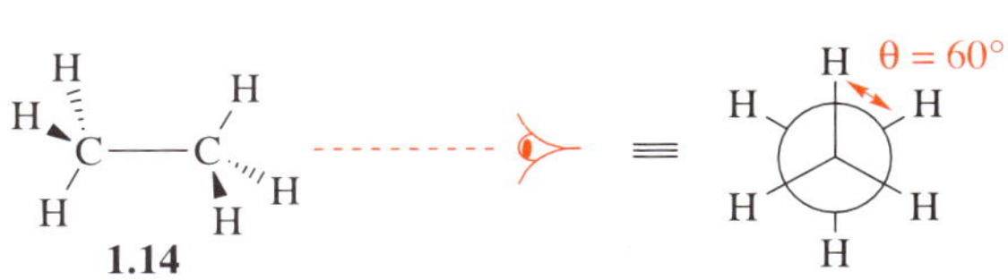

1.14

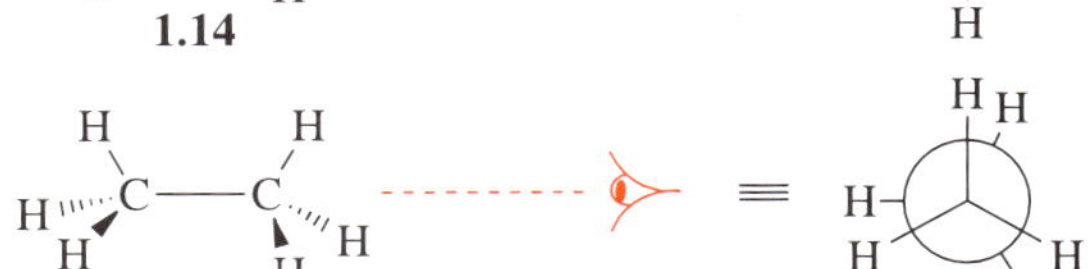

1.15

When these ideas are extended to butane, there are three **rotamers** about the central C–C bond which need to be considered. Not only are there the extremes of the staggered (**1.16**) and eclipsed conformations (**1.17**), in which the methyl group interactions are at a minimum and a maximum respectively, but there is also a *gauche* conformation (**1.18**) which is intermediate between these.

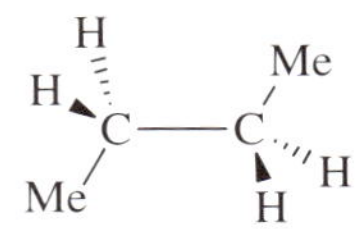

1.16

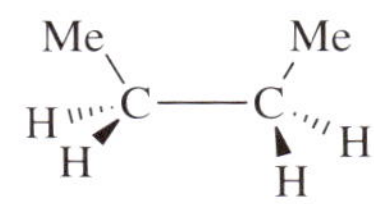

1.17

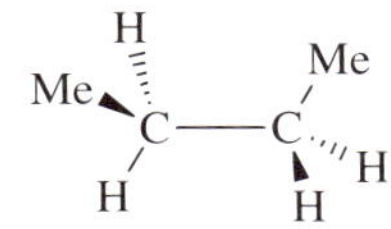

1.18

1.19 boat

1.20 chair

When the carbon chain is constrained in a cyclohexane ring, there are two extreme conformations known as the **boat** (**1.19**) and **chair** (**1.20**) forms. The former is destabilized by eclipsed interactions whilst in the latter the interactions are *gauche*. This conformation is more stable.

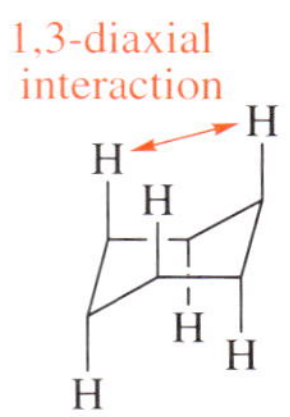

1.21

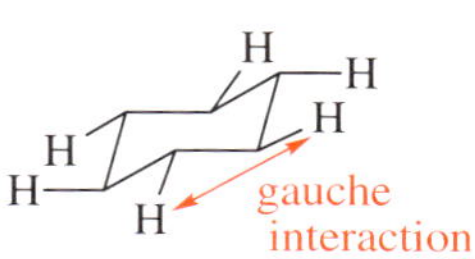

1.22

The C–H bonds on the chair cyclohexane ring are of two types. One set of six C–H bonds are parallel to the axis of the ring and are known as **axial** bonds (see **1.21**). The other set of six bonds point out of the horizontal plane of the ring and are known as the **equatorial** bonds (see **1.22**). When these C–H bonds are replaced by substituents, the substituents experience different interactions depending on their conformation. These steric relationships play an important role in the influence of the carbon skeleton of a particular compound on the reactivity of its functional groups.

In an alkene such as ethene, the presence of the π-bond prevents rotation about the C=C bond. The hydrogen atoms on the separate carbons are either *cis* or *trans* to each other. When the alkene bears substituents on the separate carbon atoms, these are *cis* or *trans* to each other. Distinct **geometric isomers** are possible. These compounds have different properties. Thus *cis*-ethenedicarboxylic acid is maleic acid (**1.23**). The carboxyl groups are close together in space and react together to form a cyclic anhydride (**1.24**). On the other hand, *trans*-ethenedicarboxylic acid is fumaric acid (**1.25**) and no such interaction is possible.

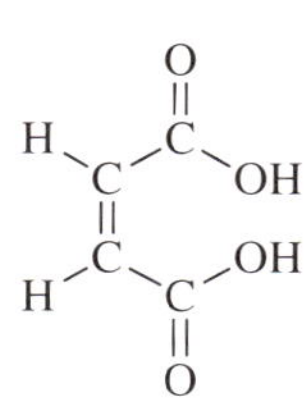

1.23

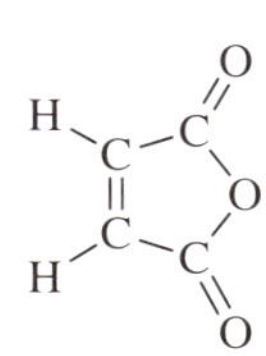

1.24

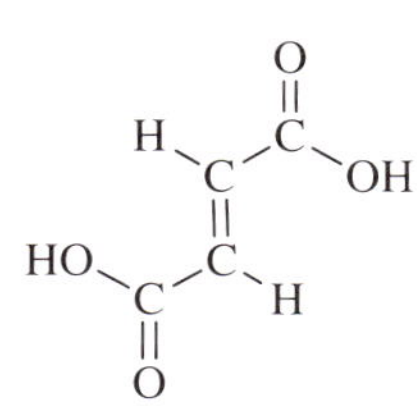

1.25

1.1.3 The Inter-relationship of Functional Groups

Functional groups may be regarded in a systematic, formal sense to be related by a series of redox and substitutive transformations. Replacement of a hydrogen atom on the carbon atom at the end of the four-carbon chain of butane (**1.26**) by a **hydroxyl** (OH) group gives the **primary alcohol** butan-1-ol (**1.27**), and when one of the methylene (CH_2) hydrogen atoms of butane (**1.26**) is replaced we obtain the **secondary alcohol** butan-2-ol (**1.28**). Replacement of the central hydrogen of the C_4 isomer 2-methylpropane (**1.29**) by an OH group gives 2-methylpropan-2-ol (**1.30**), a **tertiary alcohol**. In each of these isomeric alcohols there is a hydroxyl (OH) group conferring similar properties. However, the alcohols differ in the number of hydrogen atoms attached to the carbon atom and hence in the properties associated with these atoms. Insertion of the oxygen between the two central carbon atoms gives a further $C_4H_{10}O$ isomer, ethoxyethane (diethyl ether, **1.31**), lacking the characteristic OH of the alcohol and thus containing a different functional group, the **ether** group.

$CH_3CH_2CH_2CH_3$

1.26

$CH_3CH_2CH_2CH_2OH$

1.27

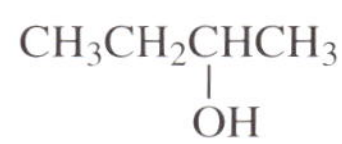

1.28

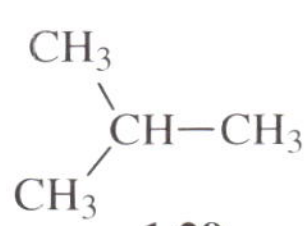

1.29

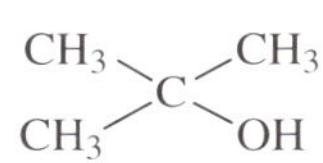

1.30

$CH_3CH_2OCH_2CH_3$

1.31

Oxidation of butan-1-ol gives butanal (**1.32**) which is characterized by a C=O, a **carbonyl** group, in this case an **aldehyde** group. Oxidation of the secondary alcohol butan-2-ol (**1.28**) gives butan-2-one (**1.33**), a **ketone**. There are many common properties of aldehydes and ketones, and others that differ because of the aldehydic C–H.

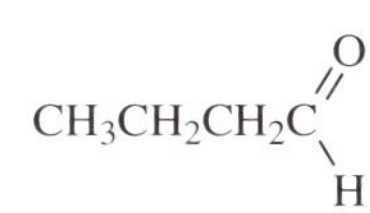

1.32

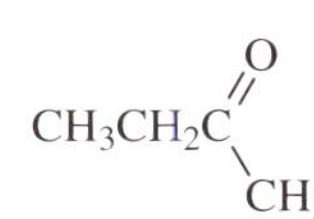

1.33

Oxidation of butanal leads to a **carboxylic acid**, butanoic acid (**1.34**). The distinctive properties of a carboxylic acid [C(=O)OH] can be considered as combining those of a carbonyl group modified by an attached hydroxyl group and those of a hydroxyl group modified by an attached carbonyl group. Replacement of the hydrogen atom of a carboxylic acid by an alkyl group gives an ester, for example ethyl acetate (ethyl ethanoate, **1.35**).

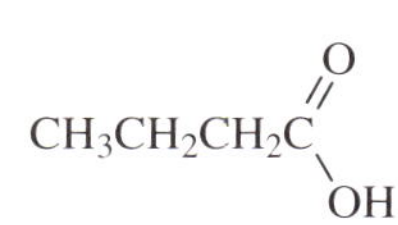

1.34

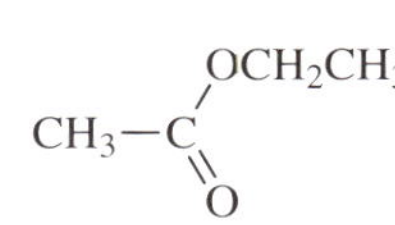

1.35

The functional group which contains two alkoxy groups attached to the same carbon atom occurs, for example, in dimethoxymethane (**1.36**), and is known as an **acetal**. A compound containing three alkoxy groups (see **1.37**) is an **ortho ester**.

If another atom such as a halogen, a sulfur or a nitrogen is substi-

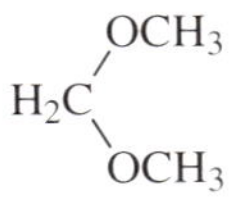
1.36

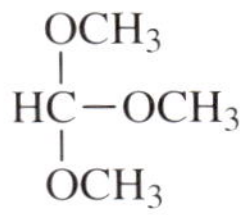
1.37

tuted in place of the hydroxyl group, further functional groups are generated (see Box 1.1).

Box 1.1 Functional Groups Derived from Alcohols and Carboxylic Acids by Substitution

R–OH	alcohol	RC(=O)–OH	carboxylic acid
R–Cl	alkyl chloride (alkyl halide)	RC(=O)–Cl	acyl chloride
R–SH	thiol (sulfane)	RC(=O)–SH	thioacid
R–NH_2	amine	RC(=O)–NH_2	amide

where R is an alkyl group.

The alkyl halides (halogenoalkanes), thiols and amines are at the same oxidation level as the alcohols, while acyl halides, thioacids and amides are similarly related to the carboxylic acids. Like oxygen, sulfur can be inserted into a chain to generate the equivalent of an ether such as the thioether.

The tervalency of nitrogen does not permit simple insertion; another group such as hydrogen or an alkyl group must be added to nitrogen, producing for example $(CH_3)_2NH$ (dimethylamine, a secondary amine) or $(CH_3)_3N$ (trimethylamine, a tertiary amine). Note that the terms primary, secondary and tertiary are used in different ways when referring to alcohols and amines.

Tertiary amines can form stable quaternary ammonium salts ($R_4N^+X^-$), while thioethers form sulfonium salts ($R_3S^+X^-$), but stable oxonium salts ($R_3O^+X^-$) are less common.

$$\underset{\textbf{1.38}}{CH_3CH_3} \longrightarrow \underset{\textbf{1.39}}{CH_2{=}CH_2} \longleftarrow \underset{\textbf{1.40}}{CH_3CH_2OH}$$

Dehydrogenation of alkanes such as ethane (**1.38**) relates them to alkenes such as ethene (ethylene, **1.39**). The same functional group may be obtained by dehydration of ethanol (**1.40**). Further dehydrogenation of ethene would generate an alkyne, ethyne (acetylene, **1.41**). In terms of oxidation level, the alkene is related to the alcohol and the alkyne is related to the ketone.

H−C≡C−H
1.41

1.1.4 Electronegativity

Having considered the oxidative and substitutive relationships between functional groups, we now consider the factors that contribute to their reactivity. The electronegativity of an element is a measure of the power of an element to attract electrons to itself in a chemical bond. It increases across a period in the Periodic Table from lithium to fluorine and decreases down a group. Electronegativity differences between atoms lead to an unequal sharing of the bonding electrons between the atoms concerned and consequently to regions of electron deficiency and electron excess in a molecule. Some electronegativities on the Pauling scale (lithium = 1 and fluorine = 4) are given in Table 1.1.

Table 1.1 Some Pauling electronegativity values

H 2.1						
Li 1.0	Mg 1.2	B 2.0	C 2.5	N 3.0	O 3.5	F 4.0
			Si 1.8	P 2.1	S 2.5	Cl 3.0
			Ge 1.8	As 2.0	Se 2.4	Br 2.8
						I 2.5

The halogens, oxygen and nitrogen are more electronegative than carbon and hence the alkyl halides, carbonyl and imine groups may be represented as in **1.42**–**1.44**, where the full-headed arrows represent the shift of an electron pair.

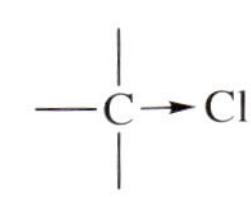
1.42

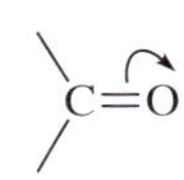
1.43

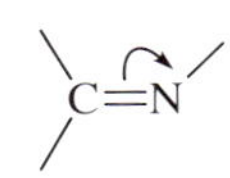
1.44

However, this is not a complete picture of the factors that contribute to the reactivity of functional groups. For example, the electronegativity difference between carbon and iodine is relatively small. In the much larger iodine atom the bonding orbitals are further from the nucleus than in chlorine and are more polarizable during the course of a reaction. These differences affect both the σ- and π-bonds. Thus many of the reactions of the alkyl halides and of carbonyl compounds may be rationalized in terms of the polarization of the bonds and the polarizability of the component atoms.

1.1.5 The Role of Lone Pairs

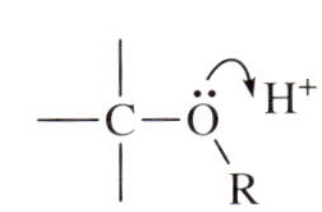
1.45

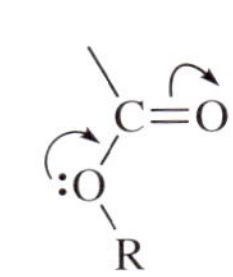
1.46

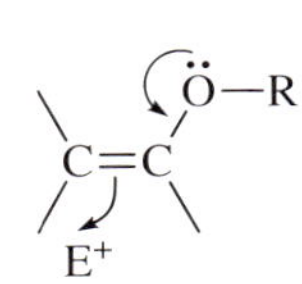
1.47

The non-bonding 'lone pairs' of electrons, particularly on oxygen and nitrogen, are far from inert and play an important role in directing the outcome of many reactions. They may accept a proton or a Lewis acid and thus increase the electron deficiency of the carbon atom to which they are attached. Secondly, they are available for donation to an attached electron-deficient carbon, and thus they may reduce the sensitivity of this carbon to nucleophilic attack. Thirdly, they are available for conjugation with the π-electrons of an alkene or arene, thus increasing its electron-rich character. These effects may be summarized in **1.45**–**1.47**. When two oxygen atoms, each possessing lone pairs, are attached to the same carbon atom, the interactions between the lone pairs become important in determining the stereochemistry of reactions.

1.1.6 Resonance

1.48 1.49

The structures of a number of compounds that contain a conjugated π-system can be written as the combination of a number of contributory valence bond structures. Thus benzene can be written as a combination of the two valence bond structures **1.48** and **1.49**. These contributory but non-isolable structures are known as resonance structures. The delocalization of the π-electrons, arising from the combination of these valence bond structures, leads to enhanced stability.

Examination of the contributory resonance structures can shed a useful light on the regions of electron deficiency and electron excess in a molecule and hence on its reactivity. The delocalization of charge through a conjugated system can give significant stabilization to reaction intermediates. An example is the delocalization of the negative charge of a carbanion adjacent to a carbonyl group (see **1.50** and **1.51**).

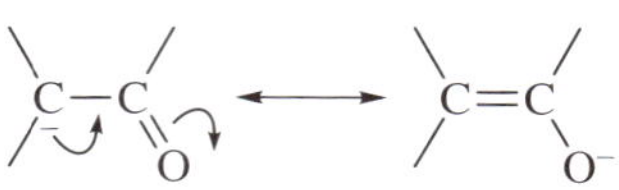

1.50 1.51

There are a number of rules that distinguish meaningful contributory resonance structures. Firstly, the atoms involved must not move between resonance structures; secondly, the same number of paired electrons should exist in each structure contributing to the resonance hybrid; and thirdly, structures that have adjacent like charges will not make a major contribution to the overall resonance hybrid, neither will those involving multiple isolated charges. Finally, it is important that the σ-bond framework, and in particular steric factors, permit a planar relationship between the contributory resonance structures.

1.1.7 Tautomerism

Compounds whose structures differ in the arrangement of hydrogen atoms and which are in rapid equilibrium are called tautomers. It is important to draw a distinction between resonance forms and tautomers. Whereas it is possible to obtain spectroscopic information on the existence of the individual tautomeric forms, resonance forms are not distinguishable. The difference can be illustrated by considering an amide (**1.52**). The resonance form (**1.53**) shows a difference in the position of charge, while the tautomer (**1.54**) shows a difference in the position of a hydrogen atom.

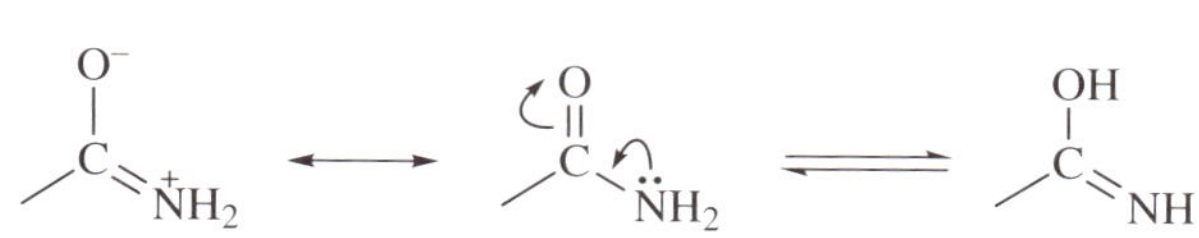

1.52 resonance 1.53 tautomerism 1.54

A number of common tautomeric relationships are shown in Box 1.2.

Box 1.2 Tautomeric Relationships

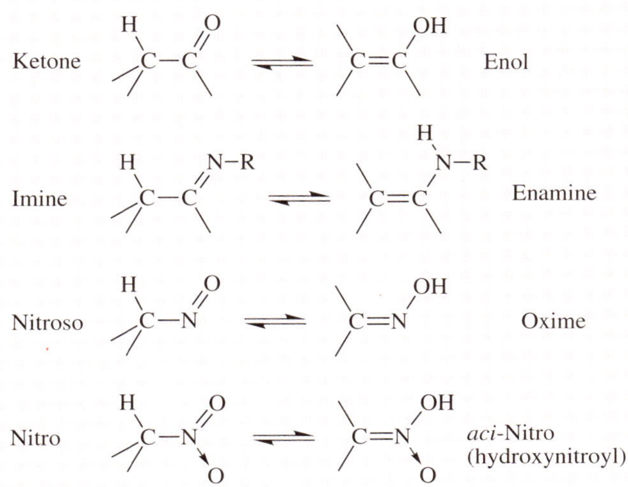

1.1.8 Naming of Compounds

Many simple common compounds are known by both a *trivial* and a *systematic* name. The systematic names are helpful in learning the structures of organic compounds, but the trivial names are often simpler and can reflect the source or dominant reactivity of the compound concerned. The systematic name for a compound has a *stem* that describes the carbon skeleton, *prefixes* and *suffixes* that indicate the functional groups, and *numbers* (locants) that define their position. Prefixes may also be added to indicate modifications to the carbon skeleton and to define the stereochemistry. A list of the more common prefixes, suffixes and stems is given in Table 1.2.

Thus the various $C_4H_{10}O$ alcohols are named as butan-1-ol [$CH_3CH_2CH_2CH_2OH$], butan-2-ol [$CH_3CH_2CH(OH)CH_3$], 2-methylpropan-1-ol [$(CH_3)_2CHCH_2OH$] and 2-methylpropan-2-ol [$(CH_3)_3COH$]. In selecting a stem, note that this includes the carbon atom of the substituent described by the suffix. For example, ethanoic acid (acetic acid) is CH_3CO_2H, not $CH_3CH_2CO_2H$ (propanoic acid). Where there is chain branching, the longest chain is selected as the stem. For example, $CH_3CH_2CH(CH_3)CH_2OH$ is named as 2-methylbutan-1-ol and not as 2-ethylpropan-1-ol.

The relative positions of substituents on an aromatic ring (*e.g.* benzene) are indicated by numbers (see **1.55**). When only two substituents

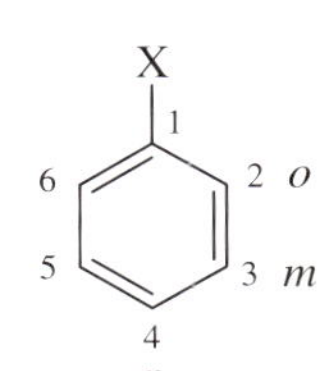

1.55

Table 1.2 Common stems, suffixes and prefixes

Examples of stems for carbon chain length

C_1	meth-	C_4	but(a)-	C_7	hept(a)-	C_{10}	dec(a)-
C_2	eth-	C_5	pent(a)-	C_8	oct(a)-		
C_3	prop(a)-	C_6	hex(a)-	C_9	non(a)-		

Examples of suffixes

-ane	alkane	-oic acid	carboxylic acid
-ene	alkene	-oate	ester or salt
-yne	alkyne	-oyl	acyl derivative
-ol	alcohol	-yl	alkyl derivative
-al	aldehyde	-amine	amine
-one	ketone	-di-, -tri-	two or three of

Examples of prefixes

cyclo-	amino-	hydroxy-
cis/trans-	bromo-	iodo-
(±)-	chloro-	methyl-
	fluoro-	nitro-

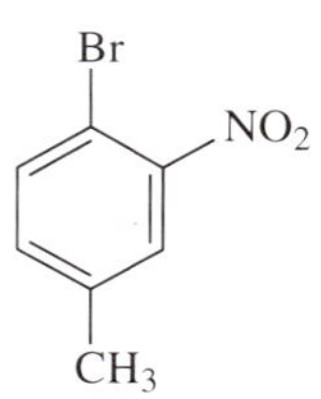

1.56

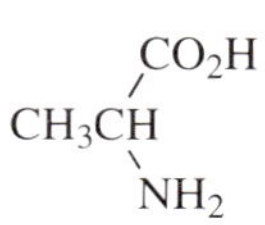

1.57

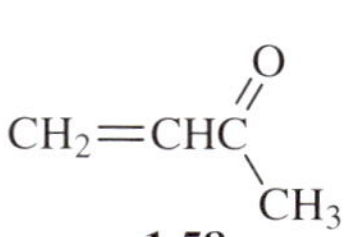

1.58

are present, *o*- (*ortho*), *m*- (*meta*) and *p*- (*para*) are sometimes used in place of 1,2-, 1,3- and 1,4-, respectively. These may also be used to relate two particular substituents in a polyfunctional molecule. The lowest numbers possible are given to the substituents (*e.g.* 1-bromo-4-methyl-2-nitrobenzene, **1.56**) and the substituents are listed in alphabetical order.

The reactivity of the positions adjacent to a functional group is often modified by the functional group. Specific names are given to these positions. The position adjacent to an alkene is known as the **allylic** position, whilst that adjacent to a benzene ring is known as the **benzylic** position. In more general cases the Greek letters α, β and γ are used to designate positions adjacent to and progressively further from a functional group. The ω-position is that at the end of a chain. Thus the common amino acids such as alanine (**1.57**) are α-amino acids and but-3-en-2-one (**1.58**) is an α,β-unsaturated ketone. The Greek letters α and β may also have a different stereochemical meaning, but the context usually makes this clear.

When two groups, for example two hydroxyl groups, are adjacent to each other, they are known as **vicinal** groups, whilst two groups attached to the same atom are referred to as **geminal** groups.

An asymmetric centre may be described systematically using the **sequence rules**. The atoms attached to the asymmetric centre are ranked according to their atomic number. The highest number is given the priority 'a' and the lowest 'd'. If, on viewing the carbon–d bond from the side remote from d, the sequence of the three higher atoms around the asymmetric carbon, a → b → c, is clockwise, the centre is described as *R* (right handed or *rectus*). If the order a → b → c is anticlockwise, the centre is described as *S* (left handed or *sinister*). The full implementation of these rules for the designation of stereochemistry, including that of alkenes, is described in books on stereochemistry.

In a number of cases, particularly with simple molecules, the commonly accepted trivial name is more clearly indicative of their properties, source and reactivity. The IUPAC rules indicate that some of these trivial names are preferred and they are in current common usage in the scientific literature and on the bottles found in the laboratory. However, systematic nomenclature is used for more complex structures, for indexing and for abstracting. Consequently, both have to be known. In this book we will use the common trivial names, giving where appropriate the systematic name as well.

Abbreviations are often used for parts of structures, particularly when these do not participate in a reaction. Thus the symbol R may be used for the remainder of a molecule. The abbreviation Ar may be used for an aromatic ring while Ph is used for phenyl (C_6H_5). Some common abbreviations for alkyl groups are given in Table 1.3.

Table 1.3 Common abbreviations for alkyl groups

Me	methyl	Bu	butyl
Et	ethyl	Bu^t or *t*-Bu	tertiary butyl (*tert*-butyl)
Pr	propyl	Ac	acetyl
Pr^i or iPr	isopropyl	Bz	benzoyl

Another abbreviation used in drawing structures is to draw only the bonds of the carbon framework, leaving out the atoms. Bonds between carbon and hydrogen atoms are also left out. Thus, butane is drawn as a 'zig-zag' and cyclohexane as a hexagon. Double and triple bonds are included as = or ≡. In this book, benzene rings will be drawn as cyclohexatrienes rather than with a circle, because this valence bond representation makes it much easier for the student to understand the mechanism of aromatic substitution. Furthermore, in polycyclic aromatic compounds the use of circles can be misleading.

used for an electron pair

used for a single electron

Electron movement is symbolized by a double-headed 'curly arrow' for the movement of an electron pair, and a single-headed arrow or 'fishhook' for the movement of a single electron. In representing electron movement, the arrow must start from the bond or atom that provides the electron(s) and the arrow should end where the electron movement terminates, either to form a bond or on the particular atom or group that receives the charge. Thus if the electron movement creates a bond, the arrow should terminate where the centre of the new bond will be; if a leaving group departs with the bonding electrons, the arrow should terminate on the atom receiving the charge.

1.2 Reagents and Reactions

1.2.1 Making and Breaking Bonds

The reactions of functional groups involve the making and breaking of bonds. In homolytic reactions the bonding pair of electrons is separated to generate two free radicals, whereas in heterolytic reactions the bonding pair stays with one partner. The reagents that bring about these reactions are thus involved in one- or two-electron processes.

In the laboratory, most reactions take place in solution. What may be considered to be a high-energy process of placing two bonding electrons on to one atom is influenced by the solvent. The solvent can play a major role in the stabilization of the reacting species by associating with the ions which are formed. Thus heterolytic reactions, which produce ionic species, are favoured in dipolar solvents with a high dielectric constant such as water, dimethylformamide (DMF, $HCONMe_2$) or dimethyl sulfoxide (DMSO, Me_2SO), whereas homolytic reactions are favoured by non-polar solvents such as petroleum ether or carbon tetrachloride.

1.2.2 Nucleophiles and Electrophiles

The reagents that bring about heterolytic reactions may be classified as **nucleophiles** and **electrophiles**. Nucleophiles are electron-rich, sometimes anionic, reagents which participate in reactions at centres of electron deficiency in a molecule. A nucleophile forms a bond to the electron-deficient centre by donating both bonding electrons. On the other hand, electrophiles are electron-deficient, sometimes cationic, reagents that react with regions of higher electron density within a molecule. The electrophile forms a bond by accepting both bonding electrons from the other component of the reaction. In Table 1.4, nucleophiles and electrophiles are listed in terms of their position in the Periodic Table.

There are some gaps in terms of common reactive species. For example, although such species or their equivalents can be generated, simple electrophiles based on oxygen (OH^+) and amide nitrogen (NH_2^+) are not commonly used.

1.2.3 Radical Reagents

Free radicals are atomic or molecular entities possessing an unpaired electron. They are formed by homolytic, or one-electron transfer, reac-

Table 1.4 Nucleophiles and electrophiles

	Nucleophiles	*Electrophiles*
	H^-	H^+ (H_3O^+)
Halogens	F^-, Cl^-, Br^-, I^-	(F^+), Cl^+, Br^+, I^+
Oxygen	H_2O, OH^-, OMe^-, OAc^-	(OH^+)
Sulfur	H_2S, SH^-, SR^-	SO_3H^+
Nitrogen	NH_3, NH_2^-, HNR_2	NO_2^+, NO^+
Carbon	R_3C^-, CN^-, $RC{\equiv}C^-$	R_3C^+, RCO^+

tions. The tendency for an unpaired electron to seek a partner means that, in the absence of special stabilizing features, free radicals are highly reactive species. Nevertheless, it is possible to design situations in which this reactivity can be turned to useful advantage.

Radicals may be generated by thermal means using, as initiators, compounds which possess either a weak O–O bond such as a peroxide, or which, on fragmentation, generate a stabilized radical and a strongly bonded product such as nitrogen gas. Azobisisobutyronitrile (**1.59**) falls into this class. After the loss of the nitrogen, the nitrile stabilizes the adjacent carbon radical by delocalization.

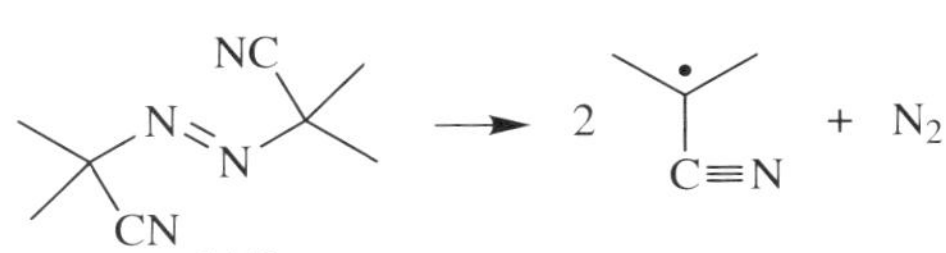

1.59

When a molecule is irradiated, particularly with ultraviolet light, some bonds within the molecule can absorb this energy and undergo homolytic cleavage to generate free radicals. These reactions include the formation of alkoxyl (RO•) radicals from alkyl hypoiodites (RO) or nitrites (RONO) or the generation of bromine atoms from bromine or *N*-bromosuccinimide.

A number of radicals may be formed by one-electron transfer redox reactions using a metal ion. These may be either oxidations in which a transition metal ion such as iron(III) accepts a single electron from the organic substrate to become iron(II), or the reaction may be a reduction in which a strongly electropositive metal such as sodium donates an electron to the substrate.

Many radical reactions differ from ionic processes in that they involve a chain of reactions. Once initiated, one radical reacts with another molecule to generate a further radical by breaking another electron pair. This in turn generates a third radical and so on, thus propagating the chain until ultimately it is terminated by the combination of two radicals.

1.2.4 Pericyclic Reactions

There are a number of concerted reactions involving cyclic transition states which are characterized by the maintenance throughout of an overlap between orbitals of the correct symmetry. These reactions are known as pericyclic reactions and the rules that govern them are known as the **Woodward–Hoffmann rules**. A typical example of a reaction of this type is the **Diels–Alder reaction** of a diene and a dienophile.

1.2.5 Acids and Bases

The Brønsted theory of acids and bases defines an acid as a proton donor and a base as a proton acceptor, *i.e.* a protic acid such as hydrochloric acid is a source of protons. Although the idea of an acidic hydrogen in organic compounds may initially be understood in terms of a carboxyl hydroxyl group, a hydrogen atom may become weakly acidic in a number of other circumstances, *e.g.* when it is attached to a carbon atom that is adjacent to a carbonyl group. On the other hand, a base such as an amine, or a carboxylate anion, is capable of accepting a proton.

$$AlCl_3 + Cl^- \longrightarrow AlCl_4^-$$

1.60

$$BF_3 + OEt_2 \longrightarrow BF_3{:}OEt_2$$

1.61

The idea of the proton as the acidic entity was extended by G. N. Lewis. Protons can accept electron pairs from bases; **Lewis acids** are generalized electron-pair acceptors. For example, aluminium trichloride (**1.60**) behaves as a Lewis acid and reacts with the chloride ion (a Lewis base). Boron trifluoride may react with the lone pair of the oxygen of diethyl ether to form boron trifluoride etherate (**1.61**). A Lewis base is an electron pair donor, in this case the ether oxygen.

The concept of **hard and soft acids and bases** is useful in classifying reagents. The nature of the outer electron shell of an atom determines its reactivity. If the electron shell is firmly bound and the orbitals are rigidly directed and of low **polarizability**, then the atom is said to be hard. If the orbitals are less rigidly held and are more polarizable, the atom is said to be softer. Hard acids tend to react with hard bases, and soft acids react with soft bases. Some typical hard and soft acids and bases are given in Table 1.5.

Table 1.5 Hard and soft acids and bases

	Hard	*Borderline*	*Soft*
Acids	H^+, Li^+, Na^+, Mg^{2+}, Ce^{4+}, Ti^{4+}, Cr^{3+}, Fe^{3+}, BF_3, $AlCl_3$, Me_3Si^+	Fe^{2+}, Cu^{2+}	Cu^+, Ag^+, Cd^{2+}, Hg^{2+}
Bases	NH_3, RNH_2, H_2O, OH^-, F^-	C_5H_5N (pyridine), Br^-	H^-, CN^-, R_2S, RS^-, I^-

Acid–Base Catalysis

Many organic reactions are subject to acid or base catalysis. For example, protonation of the oxygen atom of a carbonyl group may enhance the electron deficiency of the carbonyl carbon atom and increase its sensitivity to nucleophilic attack.

The catalyst may serve to generate the reactive species. For example, many electrophiles are generated by mineral acid or Lewis acid catalysts. Bromine reacts with iron(III) bromide to give the bromonium ion (**1.62**), whilst acetyl chloride in the presence of aluminium trichloride reacts as an acylium ion (**1.63**).

$$Br_2 + FeBr_3 \longrightarrow \underset{\textbf{1.62}}{Br^+} + FeBr_4^- \qquad CH_3C(=O)Cl + AlCl_3 \longrightarrow \underset{\textbf{1.63}}{CH_3C{\equiv}\overset{+}{O}} + AlCl_4^-$$

Base catalysis operates in a similar manner. An acidic proton of a methylene adjacent to a carbonyl group may be removed by a base to generate the reactive nucleophilic carbanion (**1.64**).

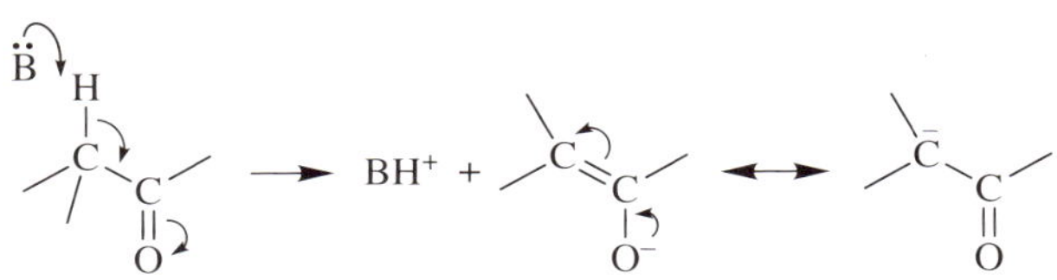

1.64

1.2.6 Reaction Types

Having considered the types of bonding, the different functional groups and the types of reagent, it is helpful to divide organic reactions into several large groups. The first group are substitution reactions in which one group directly displaces another (Scheme 1.1a). These reactions are typical of σ-bonded C–X systems. Elimination reactions (Scheme 1.1b)

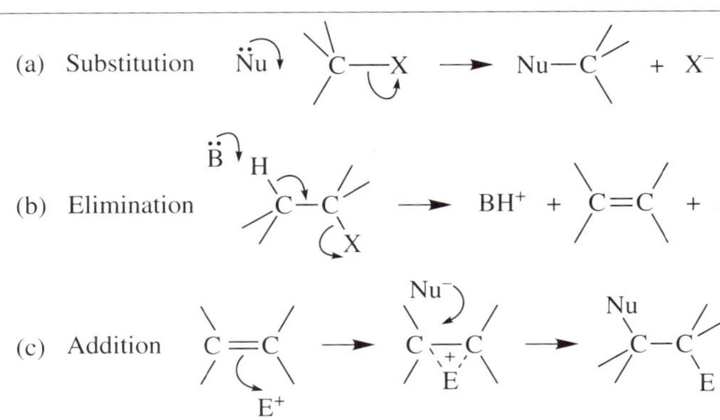

Scheme 1.1

form a second group. Elimination reactions lead to the formation of an unsaturated π-bonded system. The converse of these are **addition reactions**, in which sp or sp^2 centres are converted to sp^2 or sp^3 centres, respectively (Scheme 1.1c). These reactions are typical of π-bonds.

A number of reactions which at first sight appear to be substitution reactions, particularly at sp^2 centres, in fact proceed via an addition–elimination mechanism (*e.g.* Scheme 1.2). Oxidation and reduction reactions may often be regarded as subsets of elimination and addition reactions, respectively. Other oxidation reactions may involve the substitution of a hydrogen atom by an oxygen atom, while some reductions involve the displacement of a substituent by hydrogen (hydrogenolysis). Rearrangement reactions (Scheme 1.3) may be considered as internal substitution reactions.

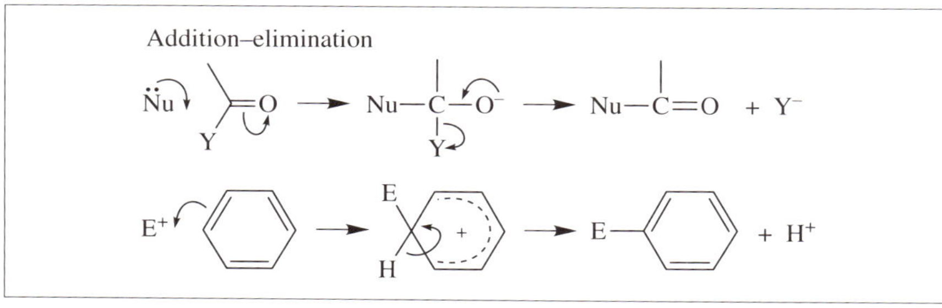

Scheme 1.2

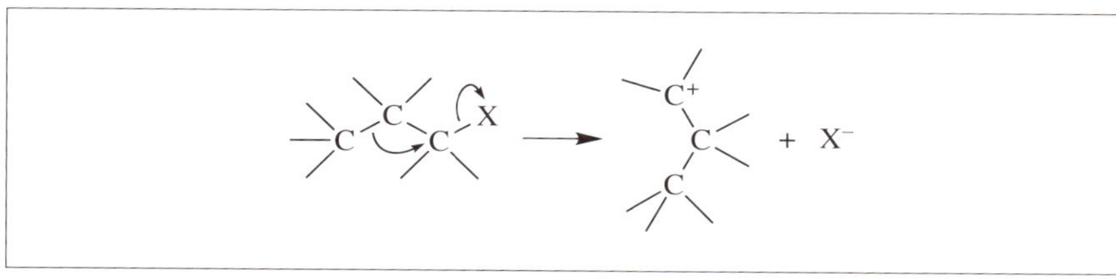

Scheme 1.3

Finally, there is the large family of industrially important polymerization reactions.

1.2.7 The Reaction Coordinate

As a reaction proceeds from starting materials to products, it is possible to show the change of the free energy against the progress of the reaction (the reaction coordinate) and thus identify various stages in the reaction (see Figure 1.1). The activation energy needed to reach the transition state determines the rate of the reaction. In a multi-step process, we can often identify the rate-limiting step.

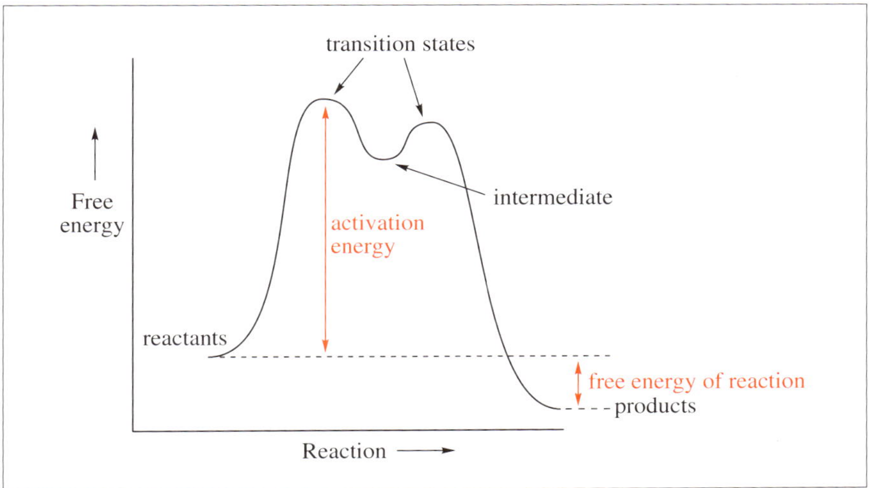

Figure 1.1 The reaction coordinate

1.2.8 Inductive and Mesomeric Effects

The electronic factors which affect the ground state of a molecule by operating through the σ-bond system, such as the electron-withdrawing effect of a halogen, are known as **inductive effects**. Electronic factors, which operate by the overlap of the p- or π-orbitals of a substituent with the π-orbitals of the rest of the molecule, are known as **mesomeric effects**. The mesomeric effect may modify the reactivity of the ground state of a molecule. Once the reaction is in progress, these effects may stabilize an intermediate by enhancing charge delocalization, leading to the **resonance stabilization** of an intermediate. The resonance stabilization of a carbanion by an adjacent carbonyl group (**1.65**) is an important example of this. A double-ended arrow (↔) is used to indicate the existence of these resonance structures.

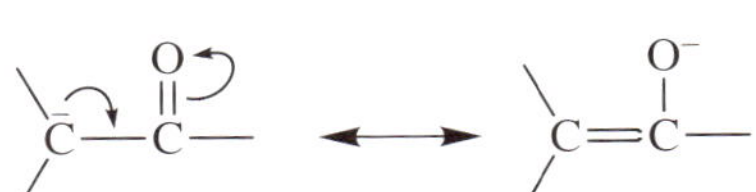

1.65

1.2.9 Kinetic *versus* Thermodynamic Control

The features that control the eventual outcome of a reaction may also be illustrated by the reaction coordinate diagram. If there are two possible reaction products with no opportunity to equilibrate between them, the product that has the lowest activation energy (and hence is formed the fastest) will be formed preferentially. The reaction is under **kinetic control**. On the other hand, if the products can equilibrate so that the most stable product is formed (*i.e.* ΔG predominates), the reaction is said to be subject to **thermodynamic control**.

1.2.10 Steric Factors

Hitherto we have considered electronic features which affect the reactivity of a functional group. However, there are a number of general steric factors which need to be considered in terms of the reaction coordinate. Although the topic will be dealt with in detail in a companion volume on stereochemistry, some general points need to be made.

The pathway of the incoming reagent approaching an sp^3 centre may strongly influence the chance of a reaction. Obstruction of this pathway by other parts of the molecule will reduce the rate of reaction by **steric hindrance**. Such steric factors can affect the face of a double bond to which groups become attached when a reagent attacks an sp^2 centre. Reagents will attack an alkene or a carbonyl group from the less-hindered face. Hydrogen bonding interactions with a reagent may also favour attack on a particular face. Since an addition leads to the conversion of an sp^2 to a more bulky sp^3 centre, the steric consequences of this have also to be considered.

Many reactions have stereochemically demanding intermediates. The requirements for the optimum overlap of the participating orbitals in many reactions may lead to specific stereochemical relationships between groups in the products of a reaction. Neighbouring functional groups may participate in many reactions, affecting not only the rate but also the products of a reaction. Each of these general stereochemical points needs to be considered for a specific reaction. Functional groups do not exist in isolation, but in real molecules in which the general pattern of reactivity may be modified by their particular environment.

1.3 Learning Organic Functional Group Chemistry

Learning organic chemistry involves rationalizing the reactivity of functional groups in a systematic way so that reactions can be seen to follow a pattern. Recognition of this pattern not only aids the understanding of organic reaction mechanisms, but also reveals those reactions which do not follow the predicted pattern and for which a special explanation must be sought.

A useful method for learning functional group chemistry is to prepare a template, such as the one illustrated in Scheme 1.4. The formula of a typical example of a compound bearing a functional group, for example a ketone such as propanone (acetone), is placed in the centre of a sheet of paper. The template containing three groups of questions is then superimposed on this. The first questions lead to the identification of the sites of electron excess and deficiency in the compound. In the case of propanone, these are the oxygen atom and the carbon atom of the carbonyl group, respectively. Secondly, the types of reaction that the compound undergoes are identified. In the example of propanone, these would be addition reactions. Thirdly, the results of using various reagents for reactions are drawn on to the sheet of paper. For propanone, these would be the products of the addition of various nucleophiles to the carbonyl group. Once a series of reactions have been outlined, it is important to rationalize them in mechanistic terms.

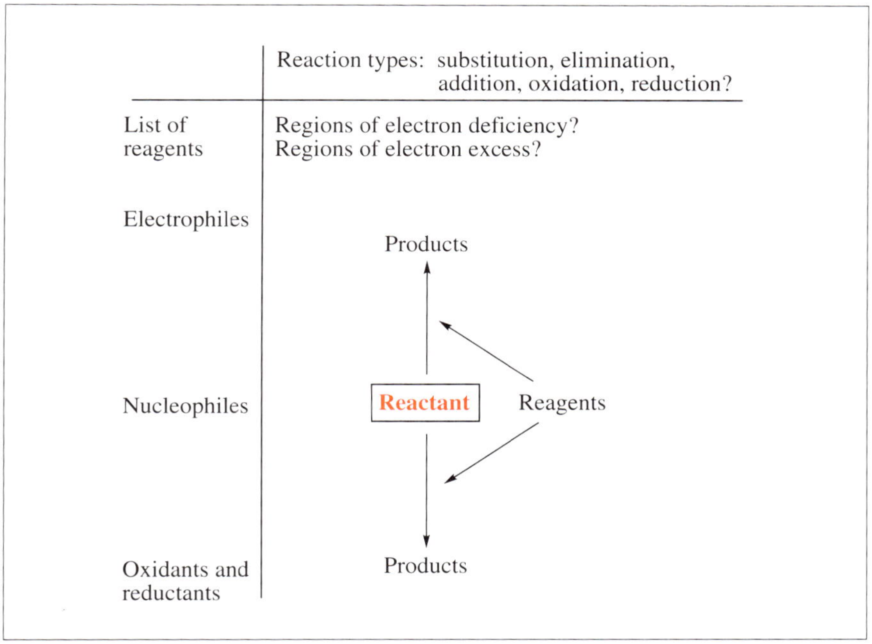

Scheme 1.4 Reaction template

A second strategy is to replace one atom by another close to it in the Periodic Table to reveal analogies and patterns of reactivity. For example, replacing a nitrogen atom by an oxygen atom may reveal analogies between the reactivity of nitriles and carbonyl compounds and between enamines and enols. It may also provide some useful ideas for synthetic methods.

A number of summary charts showing the inter-relationship of functional groups are placed at the end of the sections in Chapters 2, 3 and 4. Students should copy these and place the formulae of appropriate reagents and conditions above the arrows linking the functional groups. The creation of 'spider's webs' linking specific compounds provides a useful method of learning functional group chemistry.

Summary of Key Points

1. The 2s and 2p orbitals of carbon may be hybridized to give the tetrahedral sp^3, planar sp^2 and linear sp arrangements. Carbon may form σ- and π-bonds to other atoms. Two or more π-bonds may be conjugated. A cyclic conjugated system containing $(4n + 2)\pi$ electrons possesses a particular stability known as aromaticity.

2. Functional groups are inter-related by a series of redox and substitutive transformations. The reactions of functional groups may be determined by the electronegativity differences between the component atoms.

3. Nucleophiles are electron-rich, sometimes anionic, reagents which participate in reactions at centres of electron deficiency in a molecule.

4. Electrophiles are electron-deficient, sometimes cationic, reagents that react with regions of higher electron density within a molecule.

5. Free radicals are atomic or molecular species which possess unpaired electrons.

6. Reactions may be grouped into substitution, elimination, addition, oxidation, reduction and rearrangement reactions.

7. Acids and bases can play an important part in the catalysis of organic reactions.

8. Electronic factors which operate through the σ-bonding system are known as inductive effects whilst those operating through the π-bonding system are mesomeric effects.

9. Resonance effects may stabilize an intermediate.

10. Steric factors may influence both the products and rates of reactions.

Worked Problems

Q Identify the σ- and π-bonds in **1**.

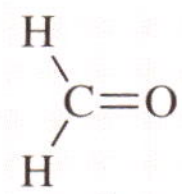

1

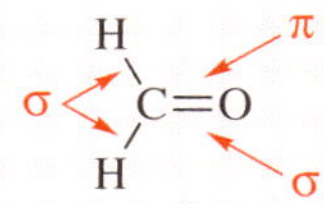

2

A See **2**. The σ-bonds are derived from the overlap between the sp^2 hybridized orbitals of the carbon and the 1s orbital of the hydrogen atoms and a 2p orbital of the oxygen. These orbitals lie between the atoms. The π-orbital lies above and below the plane of the C=O group and arise from the overlap of the 2p orbitals on the carbon and the oxygen.

Q Name compound **3**.

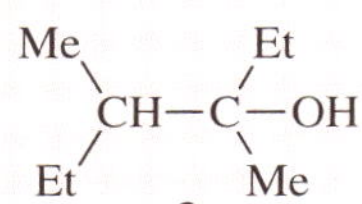

3

A The longest chain is six carbons and hence the compound is named as a substituted hexane: 3,4-dimethylhexan-3-ol.

Q Draw the preferred conformation of 1-iodo-2-phenylethane.

A See **4**. There is a staggered conformation about the ethane C–C bond with the two bulky groups *trans* to each other.

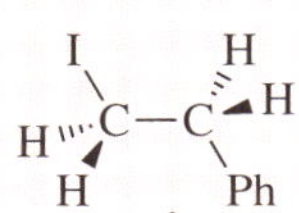

4

Q From a consideration of the Pauling electronegativity values, predict the charge distribution in the carbon–silicon bond.

A Carbon has a Pauling electronegativity of 2.5 whilst silicon is more electropositive (1.8). Consequently the charge distribution is $C^{\delta-}$–$Si^{\delta+}$.

Problems

1.1 (i) Identify the σ- and π-bonds in the following: (a) CH_3CH_2OH; (b) CH_3CHO; (c) $CH_3CH{=}CH_2$.
(ii) Identify the π-bonds in the following and comment on their structure: (d) $CH_2{=}CHCHO$; (e) benzene.

1.2 Identify the functional groups in the following structures, and indicate the regions of electron excess (δ+) and electron deficiency (δ–).

(a) 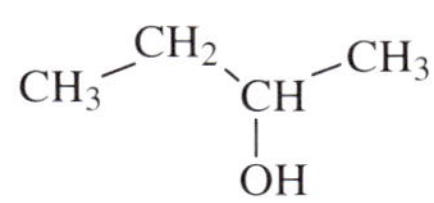

(b) 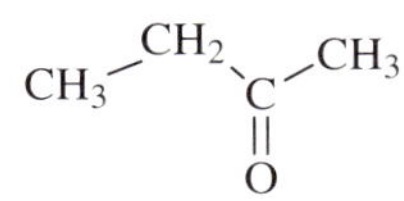

(c) 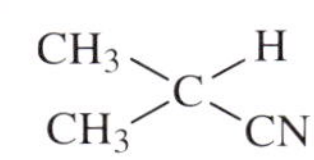

(d) $CH_3-C\equiv C-H$

(e) 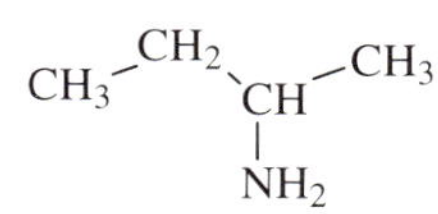

(f) 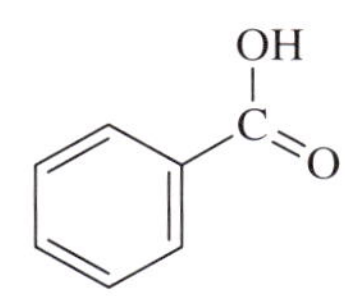

(g) 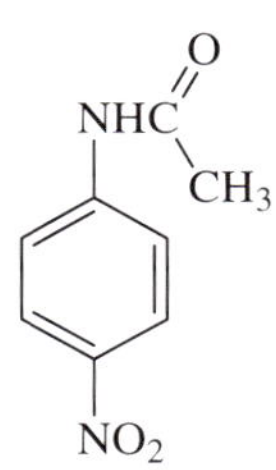

(h) 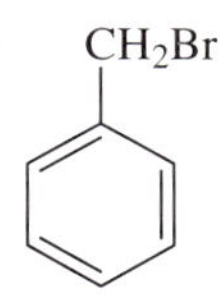

(i) 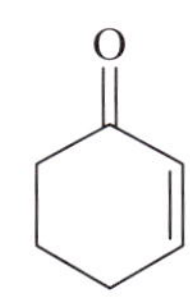

1.3 Name the following compounds:

(a) 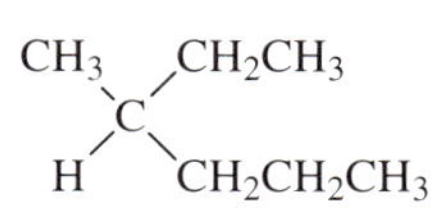

(b) 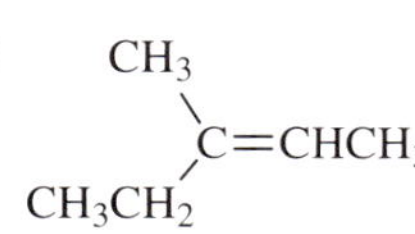

(c) 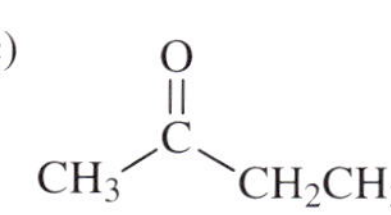

(d) 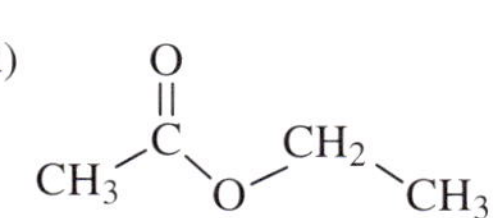

(e) 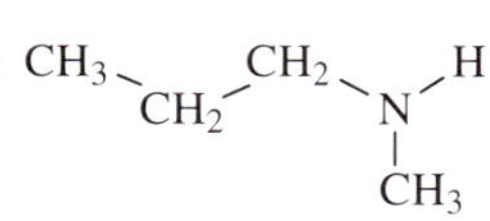

(f) 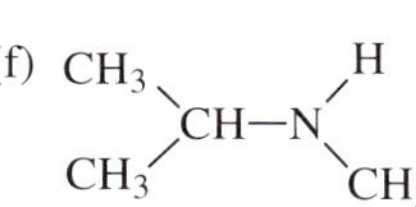

(g) 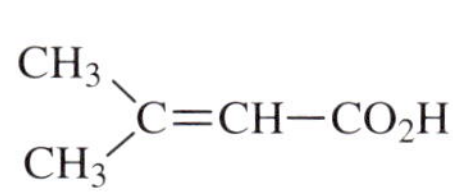

(h) 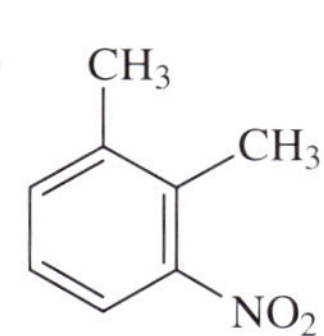

(i) 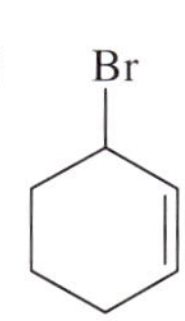

1.4 Identify the following species as nucleophiles or electrophiles: (a) Br^+; (b) Br^-; (c) CN^-; (d) NO_2^+; (e) NO_2^-; (f) NH_2^-; (g) $RC{\equiv}C^-$; (h) $MeCO^+$.

1.5 Draw resonance structures for the following:

(a) 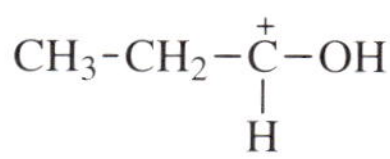

(b) 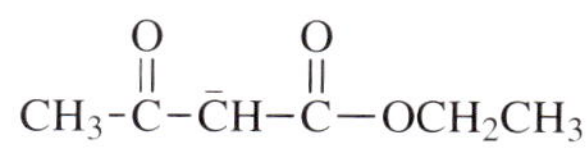

(c) 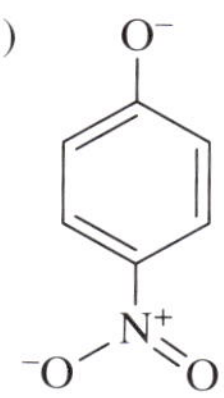

(d) 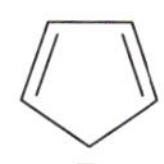

1.6 Draw the three-dimensional structures of the following, indicating the interactions that may exist: (a) *n*-butane in its staggered form; (b) *n*-butane in its eclipsed form about the 2,3-bond; (c) 1,2-dibromoethane in its most stable form; (d) cyclopropane; (e) 1,2-epoxyethane (ethylene oxide) showing the lone pairs of electrons on the oxygen; (f) *cis*- and *trans*-1,4-dimethylcyclohexane in the chair form; (g) *trans*-cyclohexane-1,2-dicarboxylic acid; (h) cyclohexene.

2 The Chemistry of the σ-Bond

Aims

This chapter of the book describes the reactions of σ-bonded functional groups. By the end of this chapter you should be able to understand:

- The difference in reactivity between a methyl, methylene and a methine hydrogen towards radical reagents
- The methods of preparing alkyl halides and the role of alkyl halides in the preparation of other functional groups
- The variation in properties between alkyl halides, alcohols, ethers, thiols and amines
- The factors which affect nucleophilic substitution of these functional groups
- The factors which affect the balance between substitution and elimination reactions
- The factors which govern the stereochemistry and regiochemistry of these reactions
- The formation of organometallic compounds from alkyl halides and their role as carbanions, thus reversing the reactive character of the carbon atom of the alkyl halide

2.1 Alkanes

2.1.1 Preparation of Alkanes

Methods for the preparation of alkanes may be divided into a number of groups. Firstly, there are methods that involve the **hydrogenolysis** of functional groups; secondly, there are methods that involve the **reduc-**

tion of unsaturated systems; and thirdly, there are methods that involve the formation of the C–C bonds of the alkane.

Hydrogenolysis of the toluene-4-sulfonate of an alcohol may be carried out with a nucleophilic hydride such as lithium aluminium hydride. There are also a series of radical methods based on the reduction of alkyl halides with tri-*n*-butyltin hydride (Bu_3SnH). Finally, the source of the hydrogen may be the electrophilic proton, exemplified by the decomposition of organometallic reagents such as the Grignard reagent with water.

The reduction of carbonyl compounds to hydrocarbons may be achieved under acidic conditions (*e.g.* the **Clemmensen reduction** with zinc and concentrated hydrochloric acid), basic conditions (*e.g.* the **Wolff–Kishner reduction** of a hydrazone with alkali) or neutral conditions (*e.g.* the catalytic reduction of thioketals with Raney nickel). The carbonyl group may represent the residue from an earlier step in the synthesis of a compound.

The catalytic reduction of alkenes and alkynes are important methods for the synthesis of alkanes. The hydroboration and hydrosilylation of alkenes are alternatives to catalytic methods. Again, both the alkene and alkyne may have played an important role in the construction of the hydrocarbon chain.

The formation of C–C bonds in hydrocarbon synthesis by various coupling reactions has a long history. While the Wurtz coupling (Scheme 2.1a) is of historical interest, modern coupling methods using organometallic reagents are more specific. Electrolytic methods (Scheme 2.1b) may lead to the generation of the free radicals involved in the coupling process.

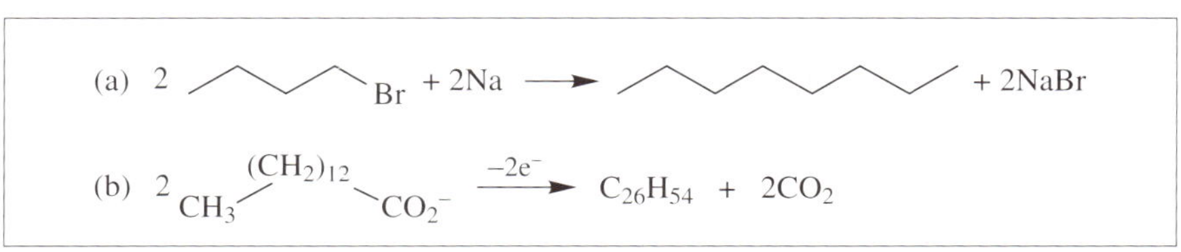

Scheme 2.1

2.1.2 Reactions of Hydrocarbons

Although the older name for the alkanes, the paraffins, arises from the Latin *parum affinis*, meaning 'little reactivity', nevertheless under the appropriate conditions the aliphatic C–H bond can be quite reactive.

Hydrogen is more electropositive than carbon and hence there is a tendency for the C–H bond to react in the sense C^-–H^+. However, very few of the reactions of alkanes are of an ionic character. The normal or straight-chain hydrocarbons are unattacked by treatment with concentrated sulfuric acid or molten sodium hydroxide. Indeed, refluxing *n*-hexane with concentrated sulfuric acid is part of the method for its purification.

The chemistry of the alkanes is dominated by the abstraction of a hydrogen atom by various free radicals. The bond dissociation energies of C–H bonds decrease in the order primary > secondary > tertiary > benzylic and allylic, and thus free radical reactions tend to occur at tertiary, benzylic or allylic centres.

The sequence of a radical chain reaction is exemplified by the photochemical chlorination of methane (Box 2.1).

Box 2.1 The Photochemical Chlorination of Methane

$Cl_2 \rightarrow Cl\bullet + Cl\bullet$	initiation
$CH_4 + Cl\bullet \rightarrow CH_3\bullet + HCl$	abstraction
$CH_3\bullet + Cl_2 \rightarrow CH_3Cl + Cl\bullet$	propagation
$Cl\bullet + Cl\bullet \rightarrow Cl_2$	termination

Overall: $CH_4 \rightarrow CH_3Cl \rightarrow CH_2Cl_2 \rightarrow CHCl_3 \rightarrow CCl_4$

In the case of the fluorination and chlorination of methane, both the hydrogen abstraction and propagation steps are exothermic, leading to a vigorous reaction. Although the abstraction of a hydrogen atom from methane by a bromine atom is endothermic, the generation of the halogen atom in the propagation step is sufficiently exothermic to allow a slower overall reaction with bromine. In the case of iodination the hydrogen atom abstraction step is sufficiently endothermic to inhibit reaction. The photochemical or radical-induced chlorination of other hydrocarbons can be quite vigorous and mixtures of mono- and polychlorinated products are obtained.

Branched chain hydrocarbons have a greater reactivity, with the relative reactivity order being tertiary C–H > secondary C–H > primary C–H. Thus, nitration of 2-methylpropane with nitric acid in a sealed tube can lead to 2-nitro-2-methylpropane.

The same reactivity order is found with oxidation. A number of these reactions may have radical character. Oxidation with chromium(VI) oxide (CrO_3) may lead to a tertiary alcohol.

When a free radical is formed and held tightly within a molecule, reactions of quite high site specificity (regioselectivity) may be observed. The free radical may be generated from another functional group in the molecule, but the result is the substitution at a centre that in a formal sense has the characteristics of a hydrocarbon. Such reactions are observed in the photolysis of nitrite esters (RONO) (the Barton reaction), the photolysis of hypochlorites (ROCl) and the oxidation of alcohols with iodine and lead(IV) acetate. These reactions generate oxygen radicals that are capable of abstracting a hydrogen atom. The radical substitution reactions then introduce a functional group at a position which is remote from conventional ionic activating groups.

2.2 Alkyl Halides

Alkyl halides (halogenoalkanes; R–X where X = F, Cl, Br or I) play an important role in many synthetic sequences. Their chemistry is dominated by the electronegativity differences between the halogen and carbon. However, this must be tempered by the ease of breaking the C–X bonds. This is reflected in the **bond energies** and in the **bond lengths** (see Table 2.1). Hence, if the rate-determining step in a reaction involves the fission of the C–X bond, fluorine will be the least reactive despite its high electronegativity.

Table 2.1 Bond dissociation energies (kJ mol^{-1}) and bond lengths (nm)

CH_3–H	426	0.109
CH_3–F	447	0.138
CH_3–Cl	339	0.178
CH_3–Br	280	0.194
CH_3–I	226	0.213

The solvation of ionic species involved in a reaction has to be considered. Furthermore, many reactions involve metal-ion catalysis in which the principles of hard and soft acids and bases come into play. Thus a large 'soft' metal may facilitate the reaction of the softer iodides, allowing a 'harder' base to react at the carbon centre. Finally, there are radical reactions based on the alkyl halides, particularly bromides and iodides, involving homolytic fission of the C–X bond.

The combination of these factors leads to the alkyl bromides and iodides being the most important alkyl halides in a synthetic context. The strength of the C–F bond and the relatively small size of the fluorine atom give rise to useful biological and industrial properties. The slow metabolic cleavage of a C–F bond can be both an advantage and a disadvantage. Alkyl fluorides have a chemistry which is somewhat different from the other alkyl halides. Highly fluorinated compounds are more stable than their highly chlorinated analogues. This is of use in the polymer industry, but it produces environmental problems in the disposal of Freons that have been used as refrigerants.

2.2.1 Methods of Preparation

Many of the convenient methods of preparing alkyl halides are based on the reactions of alcohols with reagents such as thionyl chloride and phosphorus pentachloride. These are dealt with in more detail in Section 2.3 on alcohols. The nucleophilic substitution of an alkyl methanesulfonate or toluene-4-sulfonate with a sodium or potassium halide is a useful method.

Another widely used series of reactions involves the addition of the hydrogen halides to alkenes. These are discussed in more detail later in the section on alkenes. Other methods that find some use include the decarboxylation of the dry silver salts of carboxylic acids by the halogens (the Hunsdiecker reaction).

A number of radical reactions involving substitution at an allylic position, for example with *N*-bromosuccinimide (NBS), are useful synthetic methods. NBS, in the presence of dibenzoyl peroxide as an initiator, reacts more rapidly at a secondary rather than a primary allylic position.

2.2.2 Reactions of Alkyl Halides

The reactions of alkyl halides are summarized in Box 2.2.

Box 2.2 Reactions of Alkyl Halides

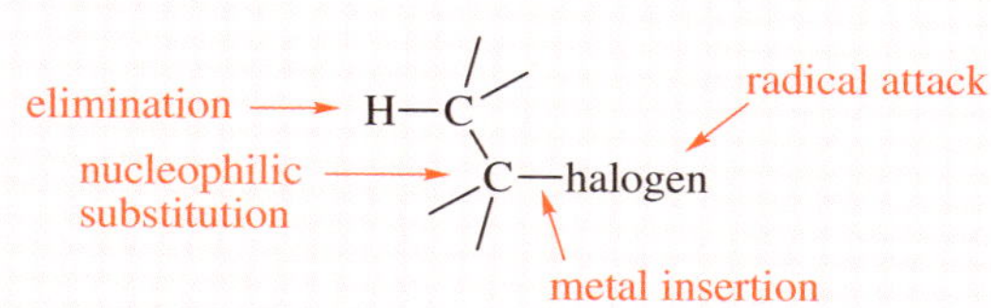

Reactions with Oxygen Nucleophiles

Many of the reactions of alkyl halides with oxygen nucleophiles represent a balance between nucleophilic substitution and elimination. These reactions may occur together, giving mixtures of products.

Two mechanistic extremes for nucleophilic substitution have been established involving unimolecular (S_N1) and bimolecular (S_N2) pathways. In the S_N1 pathway the rate-determining step is the fission of the C–X bond. This precedes attack by the nucleophile. In the S_N2 pathway, collision between the nucleophile and the alkyl halide brings about reaction and is the rate-determining step. Tertiary alkyl halides follow an S_N1 pathway while primary halides follow an S_N2 pathway. Secondary alkyl halides may follow either pathway, with the balance depending on the specific conditions. These are dealt with in more detail in books on organic reaction mechanisms, but are summarized in Scheme 2.2.

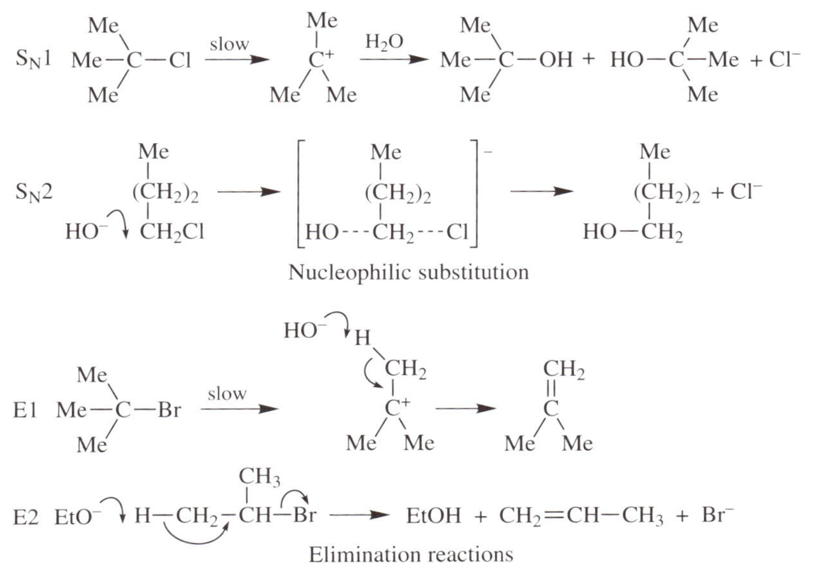

Scheme 2.2 Substitution and elimination reactions of alkyl halides

Whereas nucleophilic substitution occurs on heating with water, aqueous potassium carbonate, silver oxide or sodium acetate, elimination reactions occur on heating an alkyl halide with ethanolic potassium hydroxide. Both unimolecular (E1) and bimolecular (E2) pathways occur, the former with tertiary and the latter with primary and secondary halides. The reactions of alkyl halides with oxygen nucleophiles are summarized in Scheme 2.3.

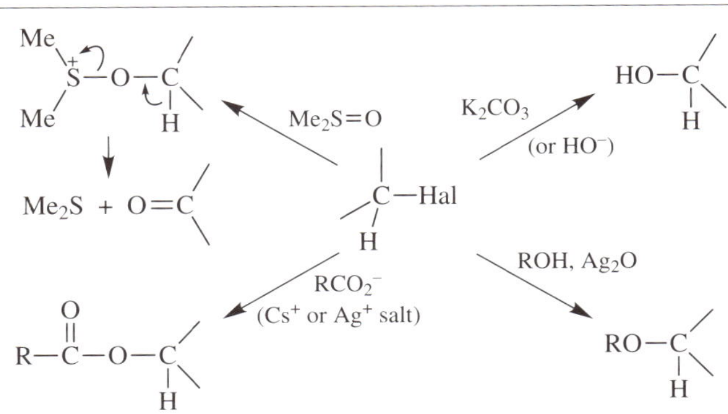

Scheme 2.3 Reactions of alkyl halides with oxygen nucleophiles

An important consequence of the bimolecular S_N2 reaction is the inversion of configuration of a chiral centre bearing the alkyl halide. As the halide departs on one face, the nucleophile approaches the other face of the carbon atom. In contrast, in the S_N1 pathway dissociation of the C–X bond takes place before the nucleophile attacks. The nucleophile may then attack from either face of the carbon atom and the stereochemical result may be racemization rather than inversion of configuration. The displacement of an alkyl halide may be accompanied by the participation of a neighbouring group. This effect can modify the stereochemistry of the reaction. A number of these reactions are facilitated by using silver salts. The silver forms an insoluble silver halide, thus driving a reaction to completion.

The reaction with dimethyl sulfoxide provides a useful method of oxidizing an alkyl halide to a ketone. Thiols are powerful nucleophiles and displace alkyl halides.

Reactions with Nitrogen Nucleophiles

The substitution reaction by ammonia and other amines can give rise to complex mixtures of primary, secondary and tertiary amines, together with quaternary ammonium salts (see Scheme 2.4). The separation of these by chemical means, as opposed to distillation or chromatography, is discussed in the section on amines.

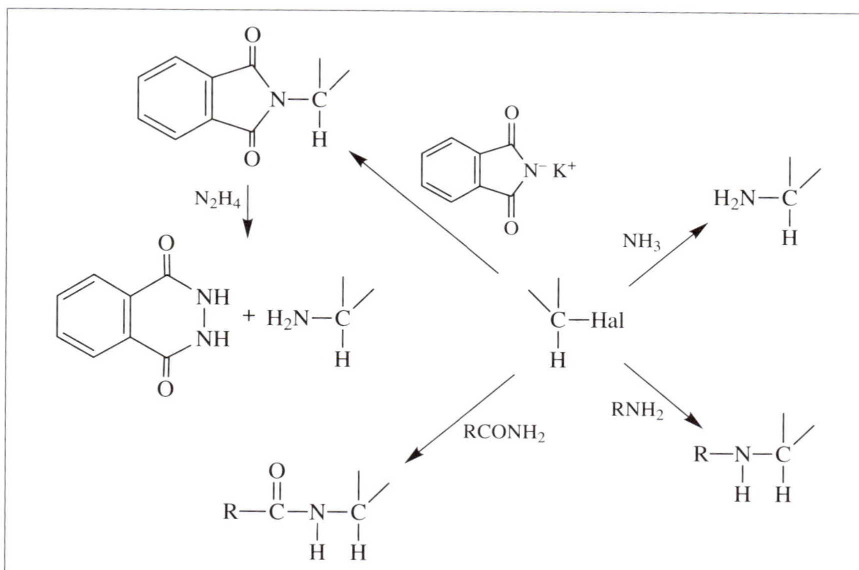

Scheme 2.4 Reactions of alkyl halides with nitrogen nucleophiles

Reactions with Carbon Nucleophiles

The substitution reactions of alkyl halides by carbon nucleophiles derived from alkynes and enolate anions provide major methods for the formation of C–C bonds (Scheme 2.5). When alkyl halides are heated with aqueous ethanolic potassium cyanide, alkyl nitriles (cyanides) are obtained, but when silver cyanide is used both cyanides (RCN) and isocyanides (RNC) are formed. A similar situation occurs with silver nitrite, which gives both a nitrite ester and a *C*-nitro compound. The synthetic value of these reactions lies in the further transformation of the products. The alkynes can be hydrated to form methylene ketones and the nitriles may be hydrolysed to form carboxylic acids.

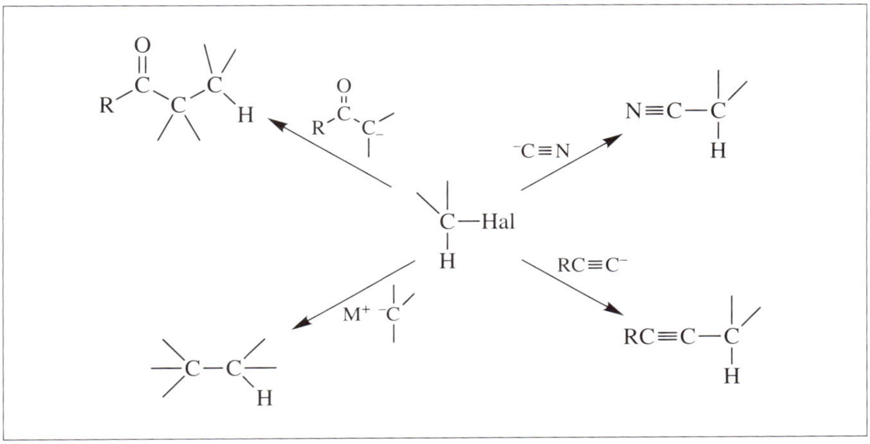

Scheme 2.5 Reactions of alkyl halides with carbon nucleophiles

Reactions involving Metal Insertion

These reactions lead to the formation of **organometallic reagents**. Since metals are more electropositive than carbon, the C–M bond is polarized in the sense C^-–M^+. Their importance in the general chemistry of the alkyl halides lies in the reversal of the reactive character of the carbon atom. It changes from being susceptible to nucleophilic attack to behaving as a nucleophile. Methyl iodide undergoes nucleophilic substitution by the hydroxide ion to form methanol in which the nucleophile has become attached to the carbon. On the other hand, treatment of methylmagnesium iodide with acid gives methane in which the electrophile is now attached to the methyl group. The best-known organometallic reagents are the **Grignard** or organomagnesium reagents. Other useful reagents are the organolithium reagents, the organozinc reagents and the organocuprates.

The Grignard reagents are prepared by treating the alkyl halide with magnesium in diethyl ether (ethoxyethane) or tetrahydrofuran. The reagent is stabilized as its diethyl ether complex. Many of the important Grignard reactions lead to the formation of a new C–C bond. These involve the **nucleophilic addition** of the reagent to a carbonyl group and are shown in Scheme 2.6. Grignard reagents also react with epoxides (oxiranes) with the formation of alcohols in which the new C–C bond is in the β-position to the hydroxyl group. The application of these reagents is described in books on synthesis.

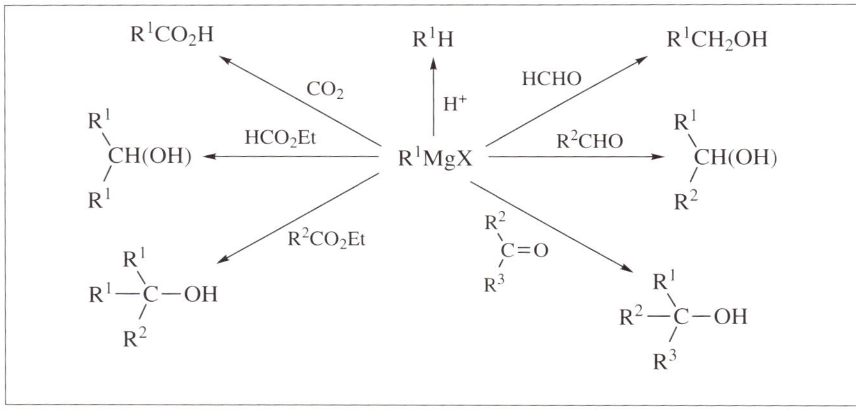

Scheme 2.6 Reactions of Grignard reagents

Radical Reactions of Alkyl Halides

The **reduction** of an alkyl halide, typically a bromide or an iodide by tri-*n*-butylstannane, is a well-established radical reaction leading to the formation of a C–H bond (Scheme 2.7a). If the carbon radical is formed in the proximity of a double bond, then a new C–C bond may be created. These reactions are useful in the formation of ring systems (Scheme 2.7b).

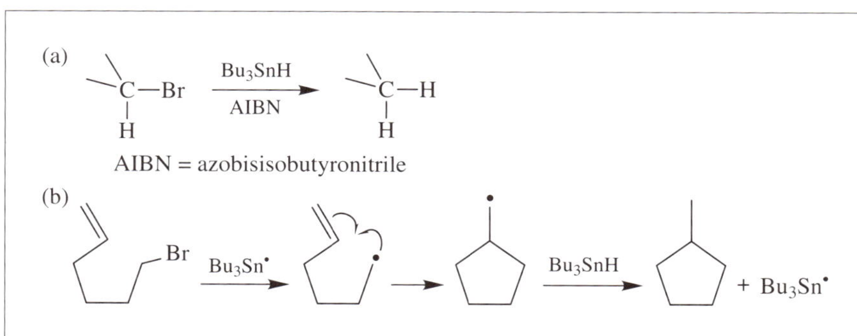

Scheme 2.7 Reductions of alkyl halides

Exercise

Copy this revision chart showing the relationship of alkyl halides to other functional groups, and fill in the relevant reagents and conditions for the inter-relationships beside the arrows.

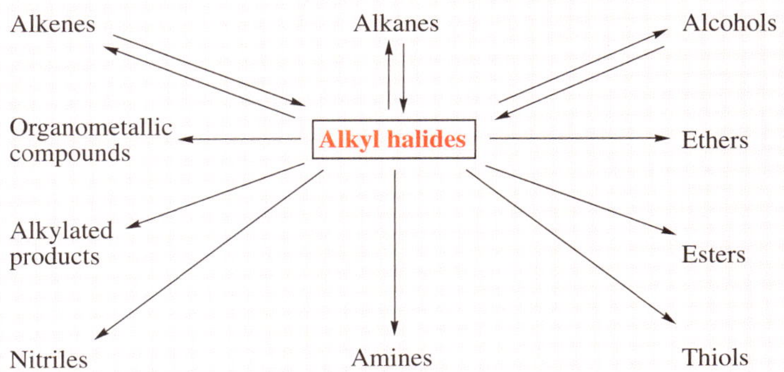

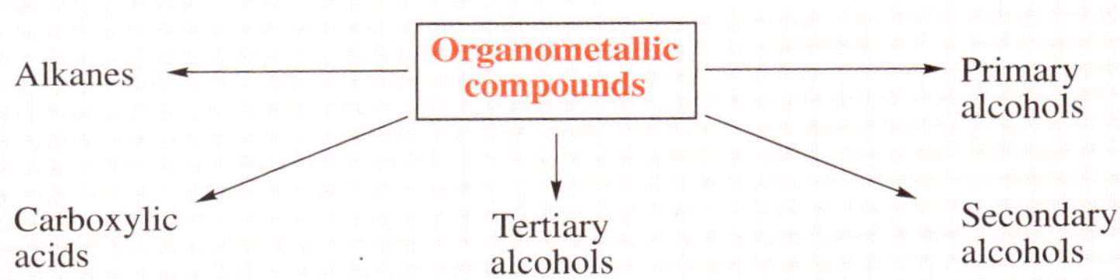

2.3 Alcohols

An alcohol (R–OH) contains a hydroxyl group in place of the hydrogen atom of an alkane. Alcohols may be primary, secondary or tertiary, depending on the number of alkyl substituents attached to the carbon atom bearing the hydroxyl group (Box 2.3).

Box 2.3 Types of Alcohol

Primary	Secondary	Tertiary
$CH_2CH_2CH_2OH$	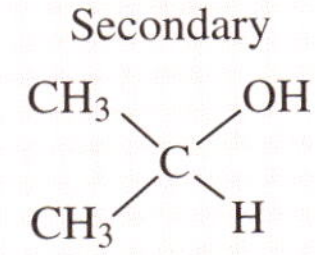	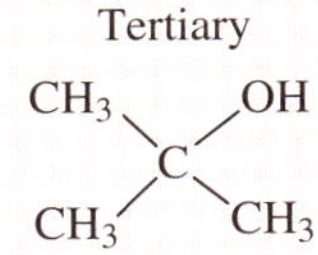
propan-1-ol *n*-propyl alcohol	propan-2-ol isopropyl alcohol	2-methylpropan-2-ol *t*-butyl alcohol

The relationship between the hydroxyl groups of a polyhydroxylic alcohol can confer special properties on these compounds. In particular, there are a number of specific reactions associated with the presence of a 1,2-diol.

2.3.1 Preparation of Alcohols

Alcohols may be prepared by the hydrolysis of alkyl halides and esters, the reduction of carbonyl compounds, the addition of carbanions to carbonyl groups and the hydration of alkenes. Although these reactions are discussed in detail in the separate sections concerning these functional groups, they are brought together here to show the inter-relationship of these functional groups with alcohols.

Alcohols may be prepared by the nucleophilic substitution of an alkyl halide with aqueous alkali or moist silver oxide (Scheme 2.8a). A second family of methods is based on the hydrolysis of esters with alkali (Scheme 2.8b).

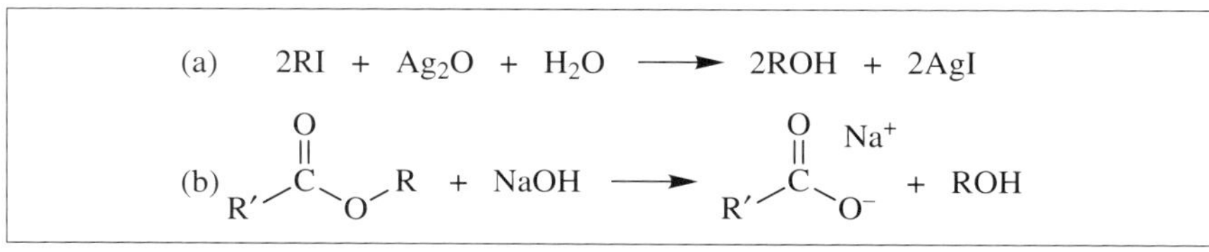

Scheme 2.8

The formation of alcohols by the reduction of aldehydes and ketones, carboxylic acids, esters and epoxides is summarized in Scheme 2.9. The change in the strength of the reducing agents, from the relatively mild sodium borohydride to the vigorous lithium aluminium hydride, reflects the difference in the electron deficiency of the carbonyl group which is being reduced.

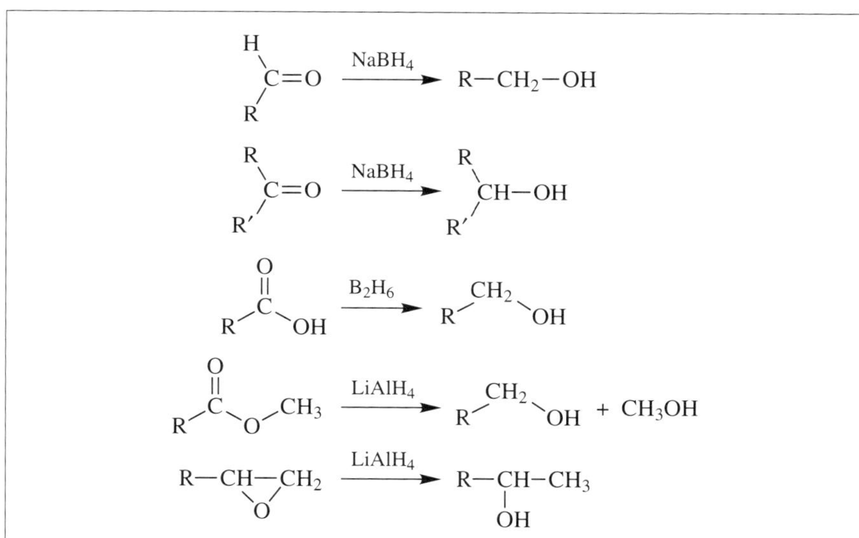

Scheme 2.9

The **hydration of alkenes** (Scheme 2.10a) is a useful way of generating alcohols. In the acid-catalysed hydration of an alkene, the initial electrophilic attack by a proton affords the more stable carbenium ion (carbocation), which is then neutralized by the attack of a water molecule. In the case of an unsymmetrical alkene this hydration follows the **Markownikoff rule**. This rule states that the electrophilic part of the addendum becomes attached to the carbon atom that bears the greater number of hydrogen atoms. This generates the more stable, more highly substituted carbenium ion.

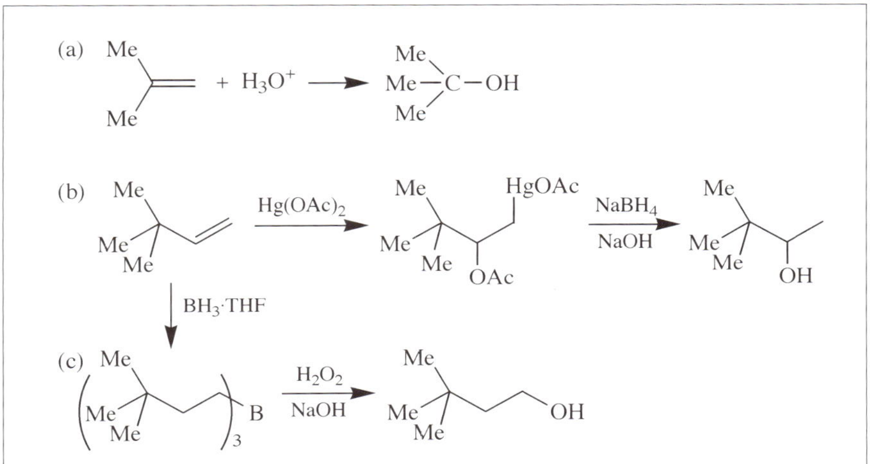

Scheme 2.10

An indirect hydration reaction may be performed using mercury(II) acetate (Scheme 2.10b). The mercury salt behaves as the electrophile, forming an organomercury intermediate. The C–Hg bond is subsequently cleaved by reduction with sodium borohydride and the alcohol is generated by hydrolysis of the acetate.

An alternative sequence involves the **hydroboration** of an alkene (Scheme 2.10c). This reaction has the overall effect of producing the **anti-Markownikoff** hydration of the double bond. A further characteristic of this pathway is that there is a *cis* relationship between the added hydrogen derived from the borane and the hydroxyl group.

2.3.2 Reactions of Alcohols

The reactions of alcohols are summarized in Box 2.4.

Box 2.4 Reactions of Alcohols

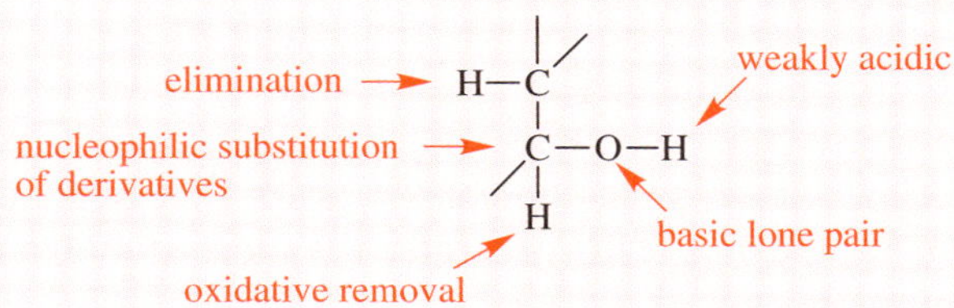

Reactions of the Hydroxyl Hydrogen Atom

The hydrogen atom of the hydroxyl group of an alcohol is weakly acidic. The hydrogen atom of the alcohol is sufficiently mobile to be exchanged rapidly with deuterium from deuterium oxide (Scheme 2.11a).

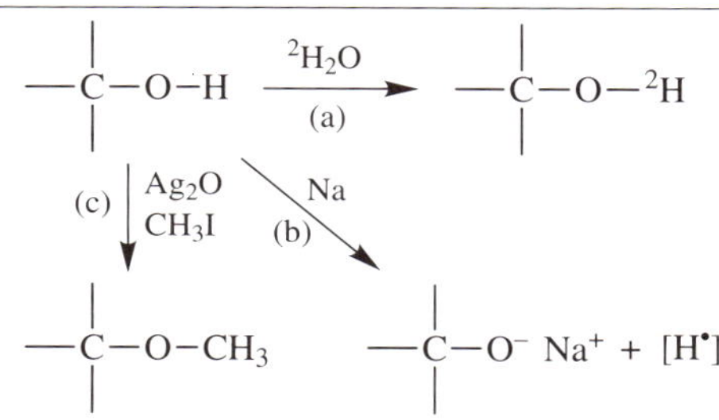

Scheme 2.11

Alcohols will form metal salts. Thus treatment with sodium, potassium or magnesium leads to the evolution of hydrogen and the formation of the **alkoxides** (Scheme 2.11b). The liberation of hydrogen in this reaction is used in the Bouveault–Blanc reduction of ketones, esters and nitro compounds.

Alkoxides are strong bases and find widespread use in the generation of carbanions by abstracting weakly acidic hydrogen atoms from, for example, the position α to a carbonyl group. The electron-donating property of the alkyl group increases the basicity of RO^- over that of the hydroxide anion (OH^-). Potassium *tert*-butoxide is a strong base but because of its steric bulk the *tert*-butoxide anion is a poor nucleophile.

Alkoxides are nucleophiles and may be used to react with alkyl halides or alkyl toluene-4-sulfonates to form ethers (Scheme 2.11c).

Reactions of Alcohols as Nucleophiles

Many of the reactions of alcohols are dominated by the availability of the lone pairs on the oxygen atom, firstly to act as a base by accepting a proton or Lewis acid and secondly to behave as a nucleophile. Protonation to form ROH_2^+ or the addition of a Lewis acid can lead to the fission of the C–O bond and the formation of a carbenium ion (carbocation). The tendency for this to occur increases as the structure of the alcohol changes from primary to secondary to tertiary. The resultant carbocation may trap a nucleophile, as in the reaction with concentrated hydrochloric acid and zinc chloride to give an alkyl halide (Scheme 2.12a). The carbocation may also react with an electron-rich system, as in the Friedel–Crafts alkylation of aromatic rings (Scheme 2.12b). Alternatively, the carbocation may lose a proton from an adjacent carbon atom to form an alkene, typified by the dehydration of alcohols with sulfuric acid or phosphoric acid (Scheme 2.12c). Finally, the carbocation may undergo rearrangements such as are found in the reactions of 2,2-dimethylpropan-1-ol (neopentyl alcohol) (Scheme 2.12d).

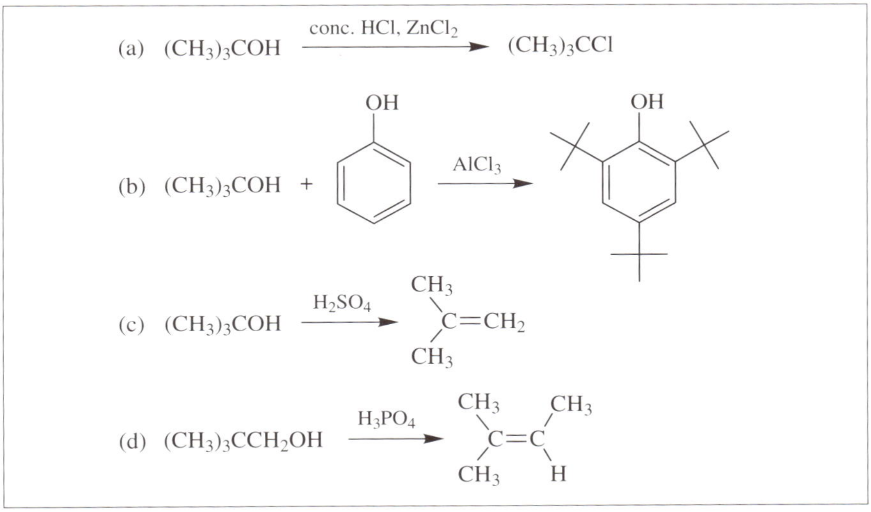

Scheme 2.12

The lone pairs may act as nucleophiles in substitution reactions of alkyl halides and sulfonates, in the solvolysis of epoxides, and in addition reactions to carbonyl groups. These reactions often proceed with acid or base catalysis.

The stereochemistry of the bimolecular reactions involves an inversion of configuration at the centre bearing the leaving group (see Scheme 2.13). When the addition takes place to an epoxide, there is a *trans* coplanar arrangement between the reacting centres: the nucleophilic oxygen of the alcohol, the carbon atom of the epoxide and the protonated oxygen of the epoxide.

The addition of an alcohol under acid-catalysed conditions to the electron-deficient carbon of a carbonyl group leads firstly to a **hemi-acetal** (Scheme 2.14a) and then to an **acetal** (Scheme 2.14b). The lone pairs of

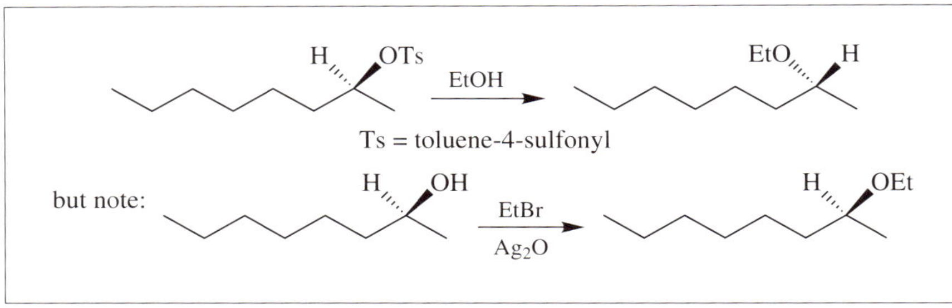

Scheme 2.13

the oxygen atom play an important role in stabilizing the intermediates in this sequence. Although the reaction proceeds smoothly with aldehydes, in the case of ketones the water which is formed has to be removed using a dehydrating agent to drive the reaction to completion.

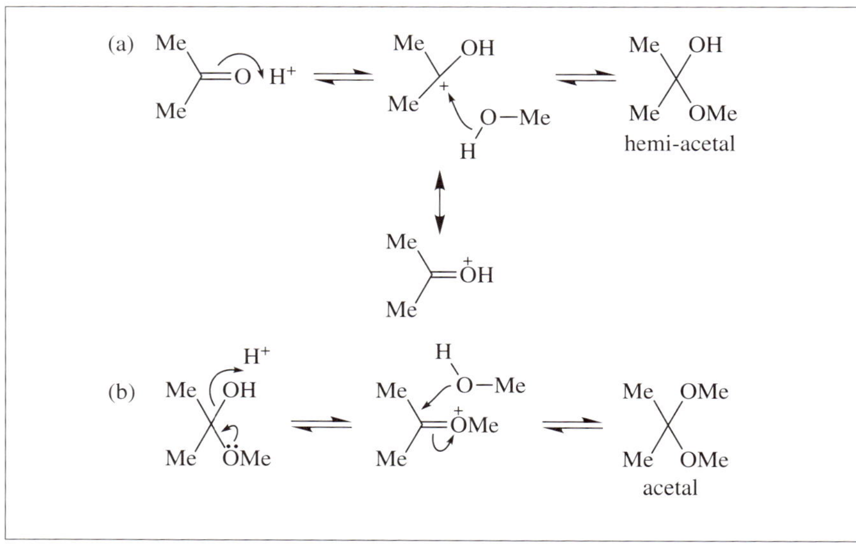

Scheme 2.14

The addition of an alcohol to the carbonyl group of a carboxylic acid in the presence of an acid catalyst leads to **ester** formation (Scheme 2.15a). The acid catalyst increases the electron deficiency of the carbonyl carbon, thus overcoming the electron-donating effect of the hydroxyl group of the acid. This enhancement of the electron deficiency of the carbonyl group of the carboxylic acid may be brought about by converting the acid to a derivative such as the anhydride or the acyl chloride. The reaction of these with alcohols leads to esters (Scheme 2.15b). Another method is to carry out the reaction of the alcohol with an acyl chloride or anhydride in the presence of a base such as pyridine, which may facilitate the removal of a proton from the alcohol.

Whereas primary and secondary alcohols form esters, these reactions

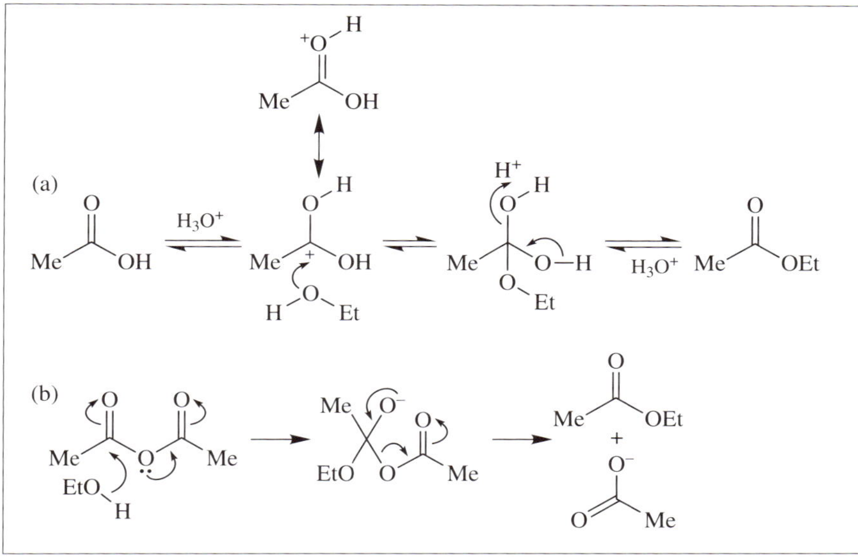

Scheme 2.15

take place far less readily with tertiary alcohols, when they may be accompanied by elimination. In such circumstances the addition of the alcohol to the highly reactive, less sterically demanding ketene (CH_2=C=O) may be useful.

Ethers, such as the benzyl, triphenylmethyl, tetrahydropyranyl and silyl ethers, acetals and esters may be used to protect the hydroxyl group against further reaction.

The electron-deficient atom with which the alcohol reacts may not be a carbon atom. It may be sulfur as in thionyl chloride ($SOCl_2$) or toluene-4-sulfonyl chloride ($MeC_6H_4SO_2Cl$), nitrogen as in nitrosyl chloride (NOCl), phosphorus as in phosphorus oxychloride ($POCl_3$) or phosphorus pentachloride (PCl_5) or chromium(VI) as in chromium trioxide (CrO_3). In each case, esters of the corresponding inorganic acids are formed. Many of these esters are very reactive and form intermediates in reaction sequences.

Conversion of Alcohols to Alkyl Halides

The reaction with thionyl chloride affords a chlorosulfite, the decomposition of which may generate an alkyl chloride by the S_Ni (substitution, nucleophilic, internal) mechanism (Scheme 2.16). This reaction, which may proceed by an ion pair, can lead to the retention of configuration of an asymmetric secondary alcohol in the conversion to the alkyl chloride. This is in contrast to the inversion of configuration found with the reaction with phosphorus pentachloride and with the nucleophilic displacement of a leaving group.

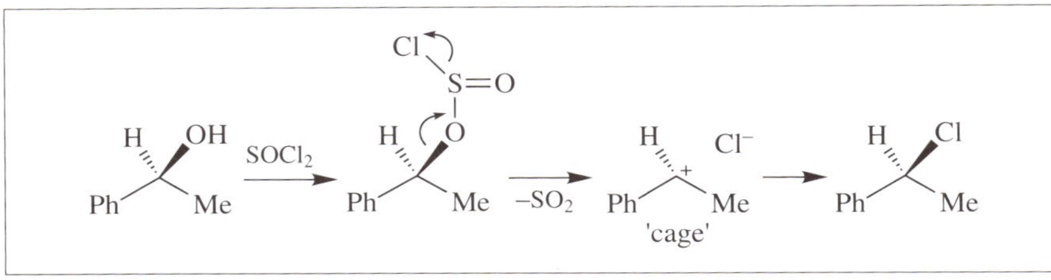

Scheme 2.16

Sulfur tetrafluoride (SF_4) or diethylaminosulfur trifluoride (Et_2NSF_3) are useful reagents for converting alcohols to fluorides. Diethylaminosulfur trifluoride will also react with aldehydes and ketones to form difluorides.

The sulfonyl halides ($ArSO_2Cl$) convert the alcohol into a **sulfonate** ($ArSO_2OR$), which is a better leaving group than the hydroxyl group. This allows a range of nucleophilic substitutions to be carried out, many of which parallel those found with alkyl halides. Alkyl halides such as iodides are formed by the nucleophilic substitution of the sulfonate by an iodide ion. The reaction in this case proceeds with inversion of configuration. Treatment of the sulfonate esters with bases such as sodium methoxide or collidine (2,4,6-trimethylpyridine), or even just heating them, can lead to the elimination of toluene-4-sulfonic acid and the formation of an alkene.

The reaction with phosphorus pentachloride, unlike thionyl chloride, proceeds more commonly with inversion of configuration.

Phosphorus Activation of Alcohols

The **phosphorus activation of alcohols** leads to substitution with inversion of configuration. The reactions are based on the high heat of formation of the P–O bond and the tendency for phosphorus to form multiple bonds with oxygen. For example, treatment of an alcohol with triphenyl phosphite and methyl iodide leads to the corresponding iodo compound (Scheme 2.17).

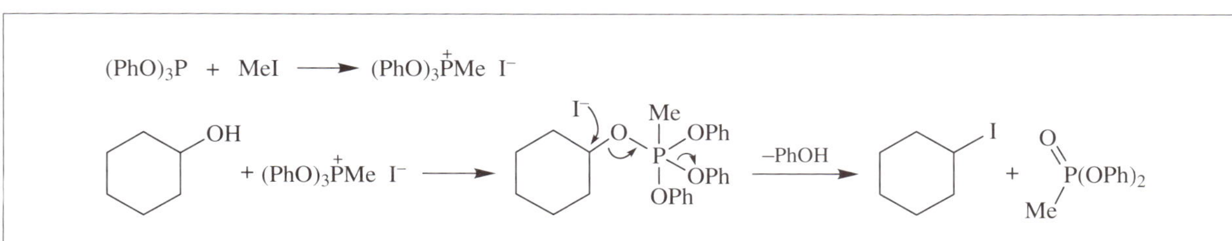

Scheme 2.17

One of the best known of these phosphorus activation reactions is the **Mitsunobu reaction** (Scheme 2.18). The phosphorus is itself first activated by reaction with diethyl azodicarboxylate (DEAD). Reaction with the alcohol then gives a phosphorus derivative which can be displaced by a variety of nucleophiles with inversion of configuration of the alcohol. If the acetate, benzoate or chloroacetate anion is used as the nucleophile, the ester of the epimeric alcohol is obtained.

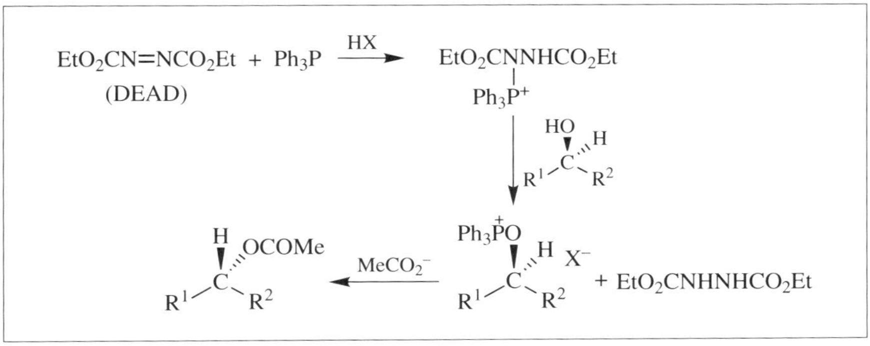

Scheme 2.18

Oxidation Reactions of Alcohols

Chromium(VI) oxide in various solvent systems provides an excellent oxidizing agent for alcohols, since it rapidly forms chromate esters which are intermediates in the oxidation of alcohols to aldehydes and ketones. The oxidation of [2-^{3}H]propan-2-ol showed a significant isotope effect when compared to propan-2-ol. Hence the abstraction of a proton by a base in the fragmentation of these esters is the rate-determining step in the reaction (Scheme 2.19).

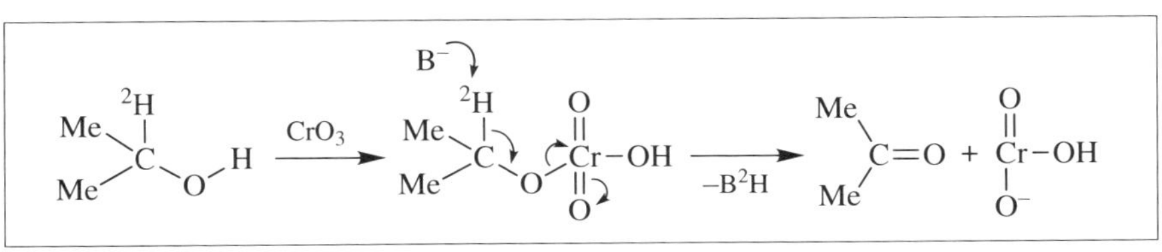

Scheme 2.19

The products of the oxidation reactions can be used to distinguish between primary, secondary and tertiary alcohols. A primary alcohol undergoes a two-stage oxidation via an aldehyde to a carboxylic acid, whilst a secondary alcohol gives a ketone. Tertiary alcohols do not react under mild conditions. Potassium permanganate can also be used to oxidize alcohols. 1,2-Diols are oxidized specifically by sodium periodate.

Remote Oxidation Reactions

A number of esters of alcohols easily undergo homolytic fission, either on heating or on irradiation with light. These reactions can lead to the generation of reactive **oxygen radicals** that may be close to other parts of a molecule. Thus in the Barton reaction, photolysis of a nitrite ester (RONO) can lead to an oxygen radical and a nitrosyl radical. The oxygen radical may remove a hydrogen atom from an appropriately placed C–H bond to form a carbon radical and an alcohol. The carbon radical may then trap the nitrosyl radical, leading to the formation of a *C*-nitroso compound and thence its tautomeric oxime.

Elimination Reactions of Alcohols

It has already been noted that a number of derivatives of alcohols, such as the sulfonate esters, when heated in the presence of a base, such as pyridine, collidine or even alumina, undergo elimination rather than substitution, so forming an alkene.

The **dehydration** of alcohols to alkenes may also occur under acid-catalysed conditions. Tertiary alcohols undergo elimination more readily than secondary alcohols, which are in turn dehydrated more easily than primary alcohols. Furthermore, where several elimination products are possible, the hydrogen atom that is lost tends to come from the carbon atom bearing the smaller number of hydrogen atoms. However, in more complex molecules this type of reaction may be accompanied by rearrangements.

Exercise

Copy this revision chart showing the relationship of alcohols to other functional groups, and fill in the relevant reagents and conditions for the inter-relationships beside the arrows.

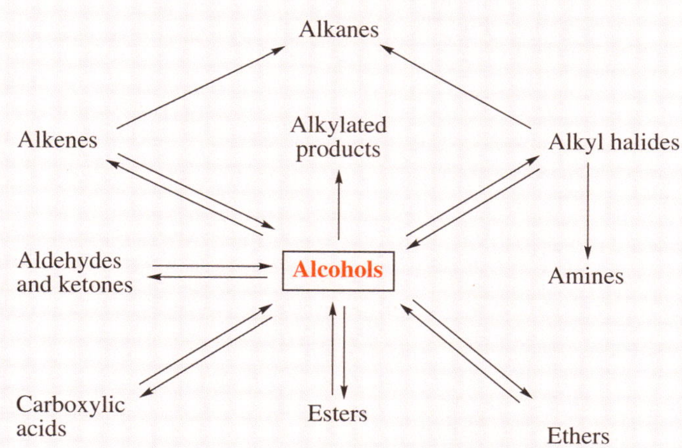

2.4 Epoxides and Ethers

Ethers (ROR′) may be considered to be the anhydrides of alcohols which are formed by the elimination of water from two alcohols (ROH and R′OH). Epoxides (oxiranes) are a sub-set of ethers and share many of their characteristics which are derived from the presence of the oxygen lone pairs. However, their much greater reactivity arising from the strained nature of the three membered ring and their wider use in synthesis means that they are best considered separately.

2.4.1 Preparation of Epoxides

Epoxides may be prepared from alkenes by the action of a peroxy acid such as *m*-chloroperbenzoic acid (Scheme 2.20a) or via the formation of a bromohydrin or iodohydrin and the treatment of this with base (Scheme 2.20b). Since the initial electrophile, the bromine or the iodine, is displaced in the second step when the epoxide is formed, the stereochemistry of this epoxidation is likely to differ from that of the reaction with peroxy acid.

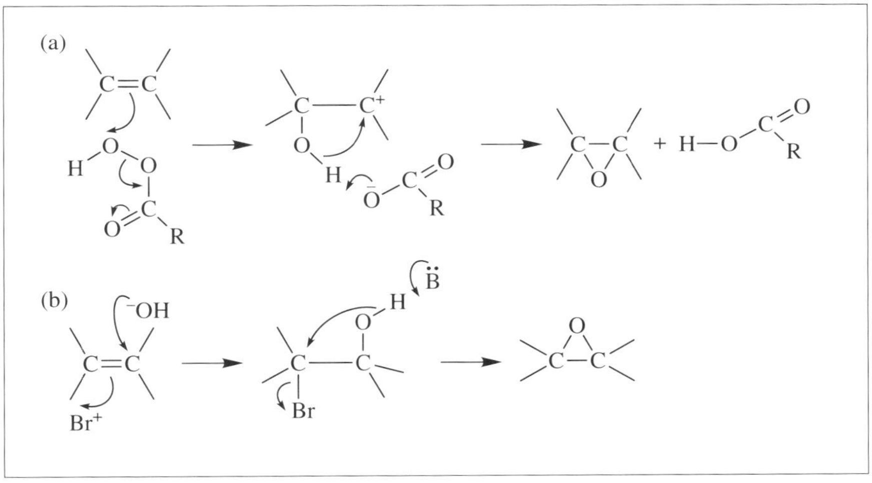

Scheme 2.20

Hydrogen peroxide or *t*-butyl hydroperoxide may be used in the presence of a catalyst such as sodium tungstate(VI) or vanadyl acetylacetonate [{MeCOCH=C(O$^-$)Me}$_2$VO] for the epoxidation of allylic alcohols. The stereochemistry of the hydroxyl group has a profound effect on the stereochemistry of epoxidation. A system which has been applied to allylic alcohols, to make optically active epoxides, utilizes titanium(IV) isopropoxide, *t*-butyl hydroperoxide and either of the enantiomeric forms of diethyl tartrate. This system forms chiral epoxides of predictable stereochemistry. When the reactivity of epoxides is combined with the synthesis of double bonds of defined geometry, this chiral step can constitute a key stage in the synthesis of optically active natural products.

The double bond of an unsaturated ketone is deactivated towards electrophilic attack by the electron-withdrawing carbonyl group but is susceptible to nucleophilic addition. Base-catalysed epoxidation with alkaline hydrogen peroxide provides a useful method for the synthesis of α,β-epoxy ketones.

The reaction of sulfur ylides (*e.g.* $Me_2S^+–CH_2^-$) with carbonyl compounds is a useful route to epoxides.

2.4.2 Chemistry of Epoxides

The chemistry of epoxides is dominated by the basicity of the oxygen lone pairs and the release of ring strain as the three-membered ring opens. Many of the reactions of epoxides are acid or Lewis acid catalysed. The catalyst may coordinate with the oxygen, increasing the polarity of the C–O bond and the sensitivity of the carbon atom to reaction with a nucleophile. The reactivity of epoxides are summarized in Box 2.5.

Box 2.5 Reactivity of Epoxides

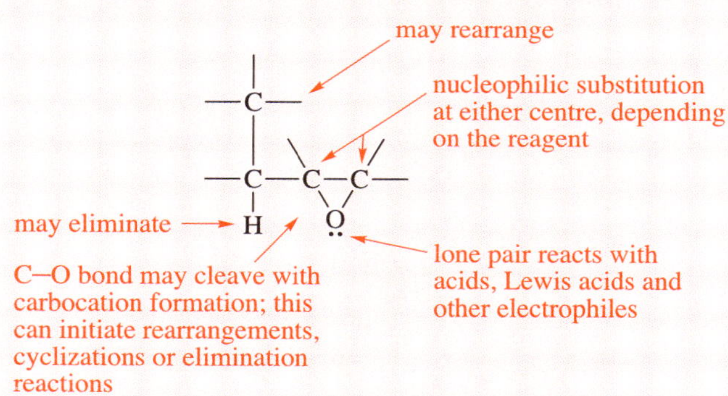

Cleavage Reactions of Epoxides

Ring opening can occur in neutral, basic or acidic media. In acidic media such as hydrochloric acid, protonation of the epoxide precedes nucleophilic attack by the halide ion or water. In general, the nucleophile then attacks from the rear of the epoxide carbon, resulting in an inversion of configuration at this centre.

With unsymmetrical epoxides, the site of the nucleophilic attack is governed by both the structure of the epoxide and the reaction conditions. In neutral or basic conditions, attack by the nucleophile at the sterically less-hindered site may predominate, but under acidic conditions there is a greater tendency for nucleophilic attack to take place at the carbon atom which can more easily accommodate a developing positive charge in the transition state.

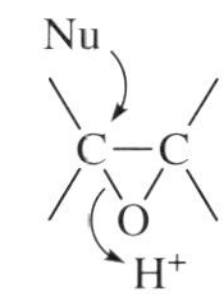

2.1

The requirement that the reacting centres must lie in the same plane for the most efficient orbital overlap leads to a *trans* relationship between the groups (see **2.1**). In a rigid cyclohexane epoxide, ring opening usually occurs to give *trans* diaxial products, although this may be influenced by neighbouring group participation from an adjacent hydroxyl group. Some examples are set out in Scheme 2.21.

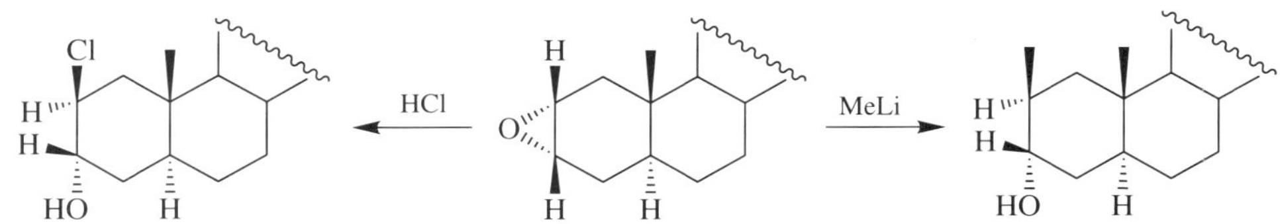

Scheme 2.21

The ring opening of epoxides with carbon nucleophiles represents a useful way of making C–C bonds. Grignard, organolithium and organocopper reagents and alkali metal acetylides have all been used for this purpose. This type of reaction has been used to form carbocyclic systems.

Epoxides may undergo rearrangement in the presence of protic or Lewis acids to give carbonyl compounds. However, the nature of the products may depend quite subtly on the reaction conditions. For example, 1-methylcyclohexene oxide has been reported to give the ring-contracted aldehyde as the major product with lithium bromide, but with lithium perchlorate, 2-methylcyclohexanone is the major product (Scheme 2.22a). In the presence of a strong base such as lithium diethylamide, an allylic alcohol may be formed from an epoxide (Scheme 2.22b).

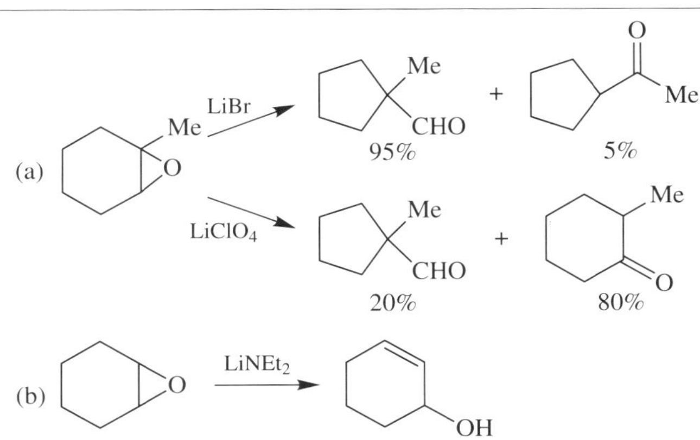

Scheme 2.22

Reduction and Oxidation of Epoxides

The reduction of epoxides to alcohols has been accomplished by a variety of reducing agents. The reduction of unsymmetrical epoxides with lithium aluminium hydride in general occurs by attack of the hydride at the least-hindered side of the epoxide to form the more highly substituted alcohol.

Organophosphorus reagents based on triphenylphosphine, or trimethylsilyl iodide, may be used to **deoxygenate** epoxides to re-form the parent alkene. Reactions based on this, or on a related scheme using the reduction of an iodohydrin, have been used in the synthesis and protection of alkenes as their epoxides (Scheme 2.23).

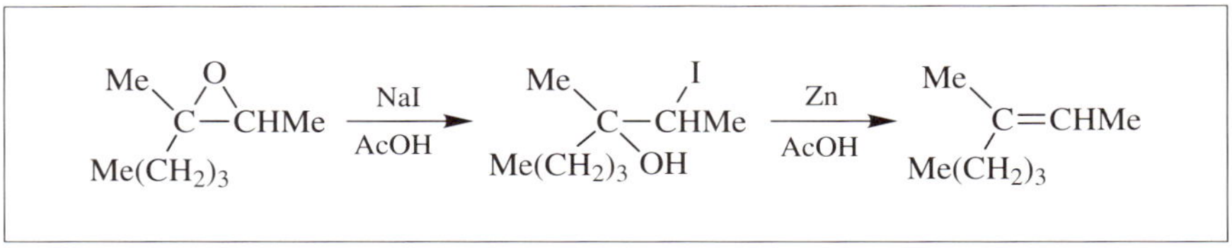

Scheme 2.23

The oxidation of epoxides by chromium(VI) oxide can be used to prepare **ketols** from trisubstituted epoxides.

Epoxides will fragment if carbanions are formed adjacent to the epoxide ring. Decomposition of the hydrazone of an epoxy ketone in the presence of base may lead to an allylic alcohol (Scheme 2.24). Since the epoxy ketone may be prepared from an unsaturated ketone, this can form part of a sequence for the 1 → 3 transposition of an oxygen function.

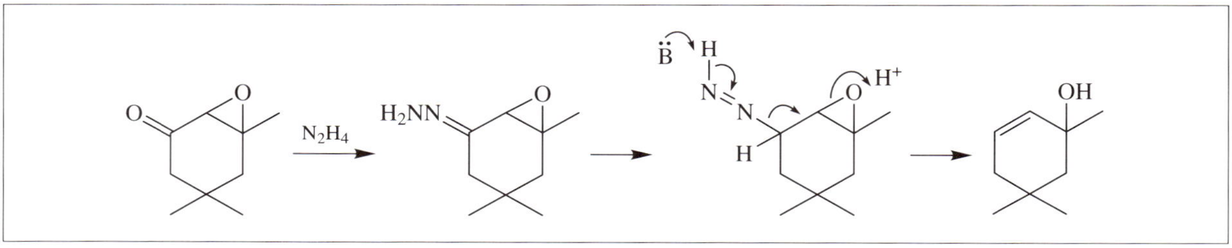

Scheme 2.24

2.4.3 Preparation of Ethers

Ethers are formed by elimination of one molecule of water from two alcohol molecules. A wide variety of dehydrating agents have been used, including sulfuric acid, phosphoric acid, potassium hydrogen sulfate and alumina.

The conversion of an alcohol to a methanesulfonate or an alkyl halide followed by nucleophilic displacement with an alkoxide is a milder procedure. Many methyl ethers have been prepared from the corresponding alcohol and methyl iodide in the presence of silver oxide (see Section 2.3.2).

2.4.4 Chemistry of Ethers

Ethers are relatively unreactive substances, which is why diethyl ether and tetrahydrofuran are widely used as solvents for organic reactions. However, the *lone pairs* on the oxygen atom are a source of reactivity. The oxygen atom may be protonated, and it reacts with Lewis acids. The increased polarity of the C–O bond then makes the neighbouring carbon atoms sensitive to nucleophilic attack.

Ethers may be cleaved with hydrogen iodide or boron tribromide to form an alcohol and an alkyl halide.

The α-position of an ether is susceptible to attack by free radicals and, in certain circumstances, by halogens. Ethers are slowly oxidized by the oxygen from air to form **peroxides**. This can be a hazard in stored bottles of ethers, particularly with the higher ethers such as di-isopropyl ether. These peroxides may be destroyed by treatment with iron(II) sulfate. Chlorine reacts with ethers, particularly in sunlight. These α-halo ethers then decompose to the aldehyde and an alcohol.

The lone pairs on the oxygen atom are effective in stabilizing an adjacent carbocation. This property accounts for the reactivity of acetals in the presence of acid. Thus chloromethyl methyl ether ($ClCH_2OMe$), obtained from formaldehyde, methanol and hydrogen chloride (Scheme 2.25a), is a powerful alkylating agent (Scheme 2.25b) and indeed it owes its high toxicity to that property. Similarly, dimethoxymethane [$(MeO)_2CH_2$] in the presence of a Lewis acid catalyst is a methylenating agent.

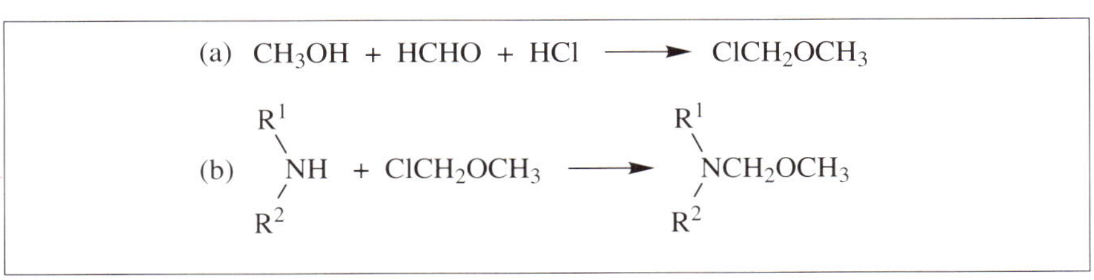

Scheme 2.25

Exercise

Copy this revision chart showing the relationship of epoxides to other functional groups, and fill in the relevant reagents and conditions for the inter-relationships beside the arrows.

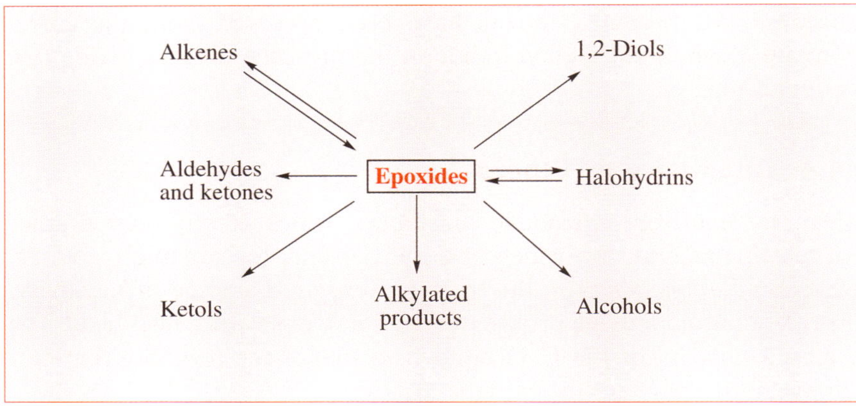

2.5 Organosulfur Compounds

Although sulfur is in the same group of the Periodic Table as oxygen, there are a number of differences between the organic chemistry of sulfur and that of oxygen. Firstly, the S–H bond is much weaker than the O–H bond. Thiols are more acidic than alcohols and the homolytic fission of the S–H bond occurs more readily. Secondly, sulfur is a larger third-row element and consequently the valence shell electrons are further from the nucleus, leading to a lower ionization potential and greater polarizability of divalent sulfur. Thus sulfur compounds are better nucleophiles than their oxygen counterparts. In hard and soft acid and base terms, sulfur is softer than oxygen. Thirdly, metal salts which readily react with sulfur compounds (*e.g.* salts of mercury and nickel) are generally different from those which react with oxygen compounds. Fourthly, sulfur has the ability to expand its valency shell and to form stable sulfonium salts. Furthermore, the availability of relatively low-lying unoccupied orbitals in divalent sulfur compounds means that sulfur can stabilize α-carbanions in a way that is not open to oxygen. Finally, the strength of the S–O bond is greater than that of an O–O bond, leading to a range of S–O derivatives. However, double bonds to carbon are more stable with oxygen (2p–2p π-bonding) than with sulfur (3p–2p π-bonding). Although sulfur-containing compounds have many applications in synthesis, their use is often restricted by the unpleasant smell associated with the more volatile compounds.

2.5.1 Preparation of Thiols

Although thiols may be prepared by the nucleophilic displacement of an alkyl halide, the resultant thiol may react further to give a thioether. Consequently, methods have been introduced which reduce the nucleophilicity of the sulfur and thus the tendency for this second step to occur. For example, reaction of an alkyl halide with thiourea gives a thiouronium salt in which the positive charge of the salt adjacent to the sulfur reduces its nucleophilicity. The thiol is then released from the thiouronium salt by treatment with base (Scheme 2.26). Aromatic thiols can also be prepared by the reduction of sulfonates or arenesulfonyl chlorides.

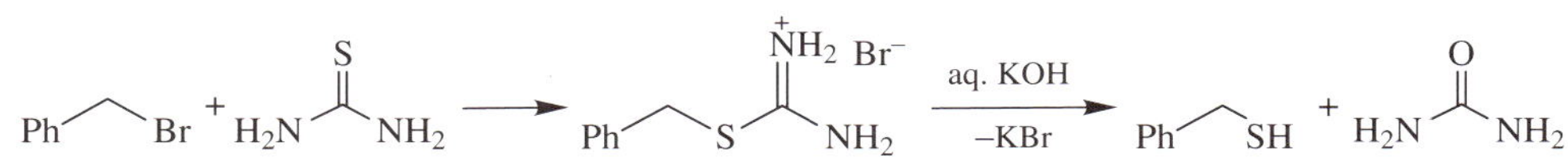

Scheme 2.26

2.5.2 Reactions of Thiols

Thiols are more acidic than the corresponding alcohols and are converted into their salts using alkali metal hydroxides or alkoxides. The thiolate salts are powerful nucleophiles, a property which has been used in the cleavage of methyl esters and methyl ethers. The combination of a hard acid and a soft base such as lithium propanethiolate ($Li^{+-}SPr$) may favour attack on the methyl group of an ester and lead to alkyl–oxygen fission of the ester, rather than the more common addition of a nucleophile to the carbonyl group and consequently acyl–oxygen fission. This is particularly useful in the hydrolysis of hindered esters.

The ease of fission of the S–H bond under **homolytic** conditions means that thiols are often used as hydrogen donors in free radical reactions.

The acid-catalysed addition of thiols to the carbonyl group of aldehydes and ketones to form **thioacetals** provides not only a useful way of protecting the carbonyl group but also of modifying the reactivity of the carbonyl carbon. The equilibrium is much more in favour of the thioacetal than the corresponding oxygen-containing acetal.

Dithioacetals of aldehydes are sources of **carbanions** and hence may be used to form new C–C bonds in reactions in which the formerly electron-deficient character of the aldehydic carbon has been reversed. The 1,3-dithianes derived from formaldehyde or a higher aldehyde may be metallated and then alkylated (Scheme 2.27). Hydrolysis of the dithioacetal is usually carried out in the presence of a thiophilic (sulfur seeking) metal salt such as a mercury salt. The insoluble sulfides cause the equilibrium to move in favour of the parent carbonyl compound.

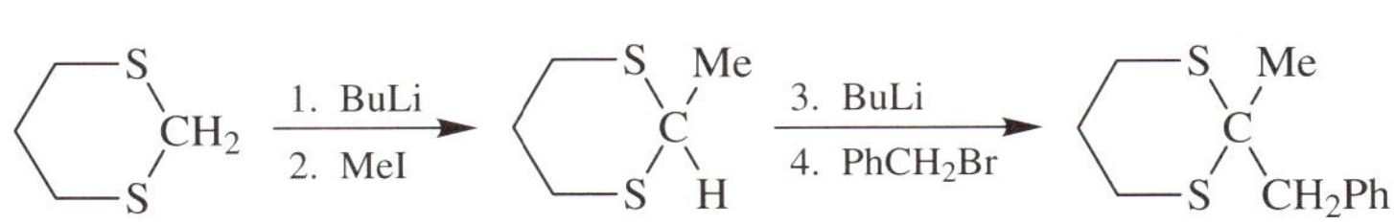

Scheme 2.27

$HSCH_2CH(NH_2)CO_2H$

2.2 Cysteine

$SCH_2CH(NH_2)CO_2H$
|
$SCH_2CH(NH_2)CO_2H$

2.3 Cystine

Hydrogenolysis of the C–S bond can be achieved both by dissolving metal systems (sodium in liquid ammonia) or by catalytic methods, particularly with a finely divided reactive form of nickel known as Raney nickel. When the latter is combined with dithioacetal formation, using either ethanedithiol or propane-1,3-dithiol, the result is a mild method for reducing a carbonyl group to a methylene group.

The oxidation of thiols in general follows a different route to that of alcohols. Thiols are easily oxidized, even by aerial oxygen, to form **disulfides**. This relationship links the amino acid cysteine (**2.2**) with its disulfide, cystine (**2.3**). These disulfide bridges are often important in holding two peptide chains together in an enzyme system. The oxidation of thioethers to form sulfoxides (R_2SO) and sulfones (R_2SO_2) may be brought about by reagents such as hydrogen peroxide, potassium permanganate or potassium persulfate. On the other hand, the oxidation of thiols to thiones (C=S) is much less common than the comparable oxidation of alcohols to ketones.

The formation of salts from sulfides is a common process. Thus reaction of dimethyl sulfide with methyl iodide gives trimethylsulfonium iodide ($Me_3S^+I^-$).

2.5.3 Sulfonium Salts

Sulfonium salts react in several ways. They may behave as a leaving group, undergoing substitution by a nucleophile or fragmenting with the formation of an alkene. However, the most important reaction of sulfonium salts involves the formation of an ylide in the presence of a base. The carbanion of this sulfur ylide is stabilized by the adjacent positively charged sulfonium ion. The reaction of the carbanion with a carbonyl group parallels that of a phosphonium ylide in the Wittig reaction. However, the decomposition of the intermediate dipolar species is different and leads to the formation of an epoxide (oxirane) rather than an alkene.

2.5.4 Sulfoxides and Sulfones

Dimethyl sulfoxide (Me_2SO), apart from being a useful dipolar aprotic solvent, has a number of reactions which reveal its behaviour as a molecule containing an $S^+–O^-$ dipole.

Alkylation may take place either on sulfur or on oxygen to form either oxosulfonium or alkoxysulfonium salts. Whereas the former are useful in carbanion chemistry, the latter can be used in oxidizing reactions. If trimethylsulfoxonium iodide is heated with a mild base in deuterium oxide, the hydrogen atoms in the salt are replaced by deuterium.

Decomposition of the deuteriated trimethylsulfoxonium iodide gives deuteriated dimethyl sulfoxide and deuteriated methyl iodide.

In its simplest application as an oxidizing agent, dimethyl sulfoxide displaces a reactive alkyl halide or sulfonate to give an alkoxysulfonium salt (Scheme 2.28a). This collapses in the presence of a mild base with the elimination of dimethyl sulfide and the formation of a new ketone or aldehyde.

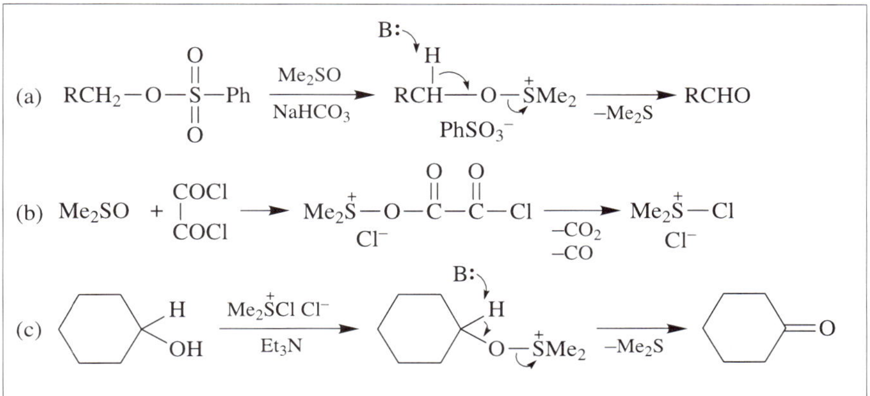

Scheme 2.28

Another method of activation is known as the Swern oxidation. Under these conditions a reactive dimethylchlorosulfonium chloride is formed from the reaction of dimethyl sulfoxide and oxalyl chloride (Scheme 2.28b). This then reacts with an alcohol to give an alkoxysulfonium salt. In the presence of a base (triethylamine) this salt fragments with the formation of a carbonyl compound (Scheme 2.28c).

The importance of sulfones (R_2SO_2) lies in the stabilization that they provide to adjacent carbanions. The carbanions are produced by treatment of the sulfone with a strong base such as butyl-lithium or lithium diethylamide. The sulfone α-carbanions may be alkylated and a variety of products may be formed. Once the sulfone has fulfilled its role in a synthetic sequence, it may be removed either by elimination as sulfur dioxide or by reductive cleavage using sodium amalgam.

2.6 Aliphatic Amines

Amines are derivatives of ammonia in which one or more of the hydrogen atoms has been replaced by an alkyl group. They are divided into primary (RNH_2), secondary (R_2NH) or tertiary amines (R_3N) according to the number of hydrogen atoms that have been replaced. The introduction of a fourth alkyl group leads to the tetra-alkyl or quaternary ammonium salts ($R_4N^+X^-$).

2.6.1 Preparation of Amines

The reaction of a primary alkyl halide with ethanolic ammonia in a sealed tube at 100 °C, or with liquid ammonia, gives a mixture of all three classes of amine together with a quaternary ammonium salt. Although mixtures of this kind may be separated by fractional distillation, this can be achieved by the **Hinsberg procedure**. This method reveals some aspects of the chemistry of amines and amides. The procedure involves converting the amine mixture to alkyl sulfonamides with benzene- or toluene-4-sulfonyl chloride. The sulfonamide of a primary amine, such as that of the ethylamine, retains an N–H group which is rendered acidic by the adjacent sulfonyl group. This sulfonamide may be extracted as a sodium salt with sodium hydroxide. The sulfonamide of a secondary amine has no acidic N–H and the lone pair of the nitrogen is conjugated to the sulfonyl group. It is no longer basic and remains in a neutral fraction. A tertiary amine such as triethylamine does not form a sulfonamide yet retains its basic character. It is therefore soluble in hydrochloric acid as a salt (Scheme 2.29).

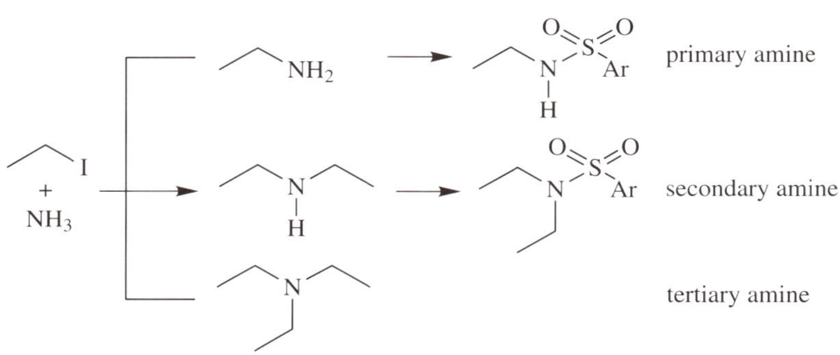

Scheme 2.29

The reduction of a number of nitrogen derivatives such as amides, oximes, alkyl cyanides or nitro compounds, particularly with lithium aluminium hydride, provides useful ways of making amines. The value of these methods lies in the fact that these derivatives allow the nitrogen to be introduced into a compound at different oxidation levels (Scheme 2.30a).

The **reductive amination** of ketones by the Leuckart method utilizes ammonium formate both as the source of the nitrogen to form an imine while the formate acts as the reductant (Scheme 2.30b).

A number of methods involve the rearrangement of carboxylic acid derivatives via nitrenes. The best known of these is the **Hofmann degradation** of amides. This involves treating an amide with bromine and alkali. The *N*-bromo compound undergoes an α-elimination in the presence

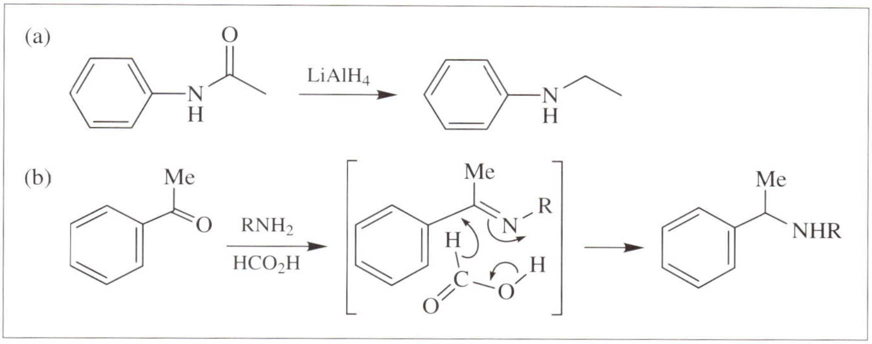

Scheme 2.30

of alkali to form a nitrene, which immediately rearranges to an isocyanate. Hydrolysis of this gives an amine (Scheme 2.31).

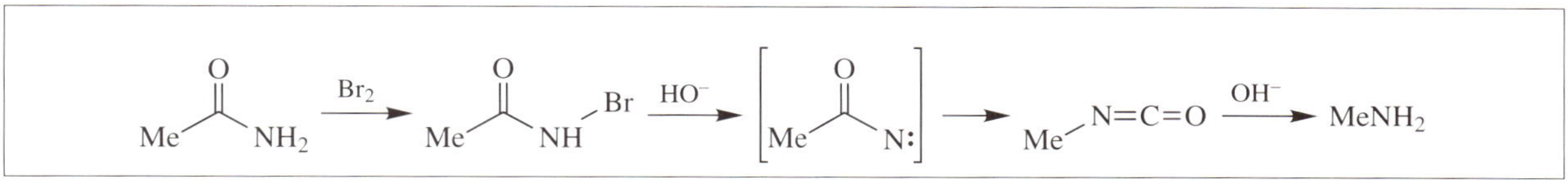

Scheme 2.31 Hofmann rearrangement

The **Beckmann rearrangement** of oximes affords amides which may be hydrolysed or reduced to form amines. The geometry of the oxime is important in directing the course of the rearrangement and hence in defining which amide might be obtained (Scheme 2.32).

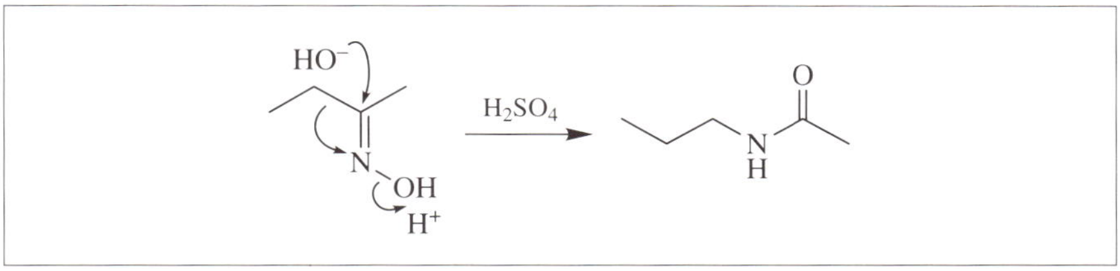

Scheme 2.32 Beckmann rearrangement

The **Gabriel method** of making primary amines (Scheme 2.33) rests on the fact that the imide hydrogen of phthalimide is rendered acidic by the two carbonyl groups. It therefore forms a potassium salt, which is a powerful nucleophile and may be alkylated. Hydrolysis of the amide generates the primary amine. The cleavage of the phthalimide may be carried out with hydrazine (H_2NNH_2).

Scheme 2.33 Gabriel amine synthesis

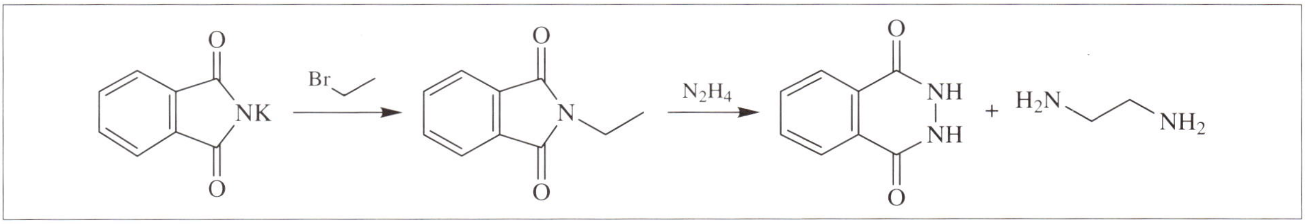

2.6.2 Reactions of Amines

The chemistry of the amines is dominated by the nucleophilicity of the lone pair on the nitrogen atom. The nitrogen can participate in both nucleophilic substitution and addition reactions. The nitrogen atom of an amine uses sp^3 hybridized orbitals in bonding. As with carbon, these are directed to the corners of a tetrahedron. Three of these orbitals, each containing one electron from the nitrogen, overlap with orbitals of a substituent whilst the fourth orbital contains the unshared lone pair of electrons from the nitrogen. Consequently, amines have a pyramidal shape with the nitrogen at one of the vertices. Under normal conditions this pyramidal shape is rapidly inverting.

Reactions with Acids

Amines are bases and form salts with acids, including carboxylic acids. They also form complexes with metal salts. As the size and number of the alkyl groups around the nitrogen increase, there is a competition between the inductive effect of the alkyl group and the steric hindrance of the group. The balance between steric and electronic factors is revealed by comparing the variation in base strength, which increases in the order ammonia (pK_b 4.7) < methylamine (pK_b 3.36) < dimethylamine (pK_b 3.29). Introduction of the third alkyl group then reduces the base strength and trimethylamine has $pK_b = 4.28$. The heat of dissociation of the amine–trimethylborane complexes ($R_3N{:}BMe_3$) rises from ammonia through methylamine to dimethylamine, whilst the trimethylamine–trimethylborane complex is about as stable as the methylamine complex. On the other hand, the triethylamine complex is quite unstable. Models show that it is possible to rotate only two of the three ethyl groups of triethylamine (**2.4**) out of the way of the trimethylborane molecule. However, quinuclidine (**2.5**) effectively possesses three ethyl groups held back in a cage and this forms an exceedingly stable addition compound with trimethylborane.

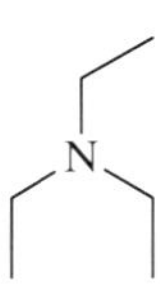

2.4

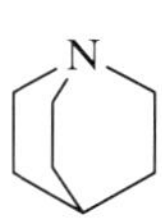

2.5

Amines in Nucleophilic Substitution Reactions

Amines can displace a halide or a derivative of an alcohol, such as the methanesulfonate or toluene-4-sulfonate, with the eventual formation of a quaternary ammonium salt. Reaction with ammonia may lead to mixtures, since the initial alkylation increases the basicity of the nitrogen, and hence the primary amine reacts more rapidly than ammonia. Further reaction, although enhanced by the increase in basicity, may be impeded by steric hindrance arising from the additional alkyl groups.

Reaction with Bases

The hydrogen of a primary or secondary amine is sufficiently acidic to form sodium or lithium **amides**. Thus treatment of diethylamine with butyl-lithium affords the strong base lithium diethylamide ($Et_2N^-Li^+$).

Amines in Nucleophilic Addition Reactions

The addition of a primary or secondary amine to a carbonyl group to form a tetrahedral intermediate is the first step in a number of reactions. Dehydration of the intermediate can give an **imine**. If, however, the carbonyl group is part of an acyl halide, ester or anhydride, the collapse of the tetrahedral intermediate may lead to an amide. These reactions are exemplified in Scheme 2.34.

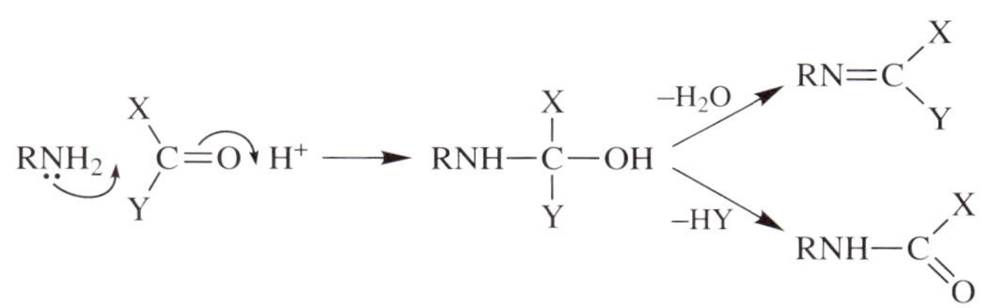

Scheme 2.34

Whereas the imines formed with aromatic aldehydes are relatively stable, those formed from aliphatic aldehydes are often unstable and undergo further reaction. If the reaction with formaldehyde is carried out with the salt of a primary or secondary amine and a ketone containing an acidic hydrogen, a new C–C bond is formed. This reaction, known as the **Mannich reaction** (Scheme 2.35), has played a useful role in many syntheses.

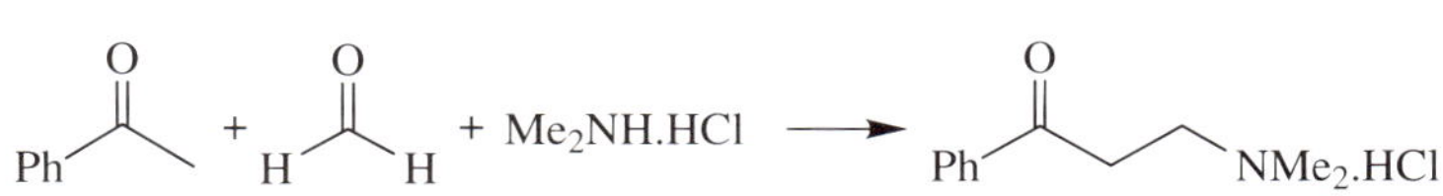

Scheme 2.35 Mannich reaction

The acetyl, benzoyl and toluene-4-sulfonyl derivatives of amines are often crystalline, and have been used to characterize the amine. The addition of an amine to an isocyanate may lead to a substituted **urea** and these have also been used to characterize the amines.

The electron-deficient atom to which the nitrogen adds may be another heteroatom, as in toluene-4-sulfonyl chloride. The product is then a **sulfonamide**. Some sulfonamides have useful biological activity as antibacterial agents.

When the electron-deficient atom is the nitrogen atom of nitrous acid or nitrosyl chloride, the products will depend on whether the amine is a primary or secondary amine. Primary aliphatic amines give rise to **diazonium salts** ($RN_2^+X^-$) which readily decompose, sometimes with rearrangement. Aromatic diazonium salts ($ArN_2^+X^-$) are very important in aromatic transformations, and are discussed later. Secondary amines, on the other hand, give rise to **nitrosamines**.

Oxidation Reactions of Amines

There are oxidation reactions in which oxygen is donated to the system (Scheme 2.36a) and there are those reactions which are dehydrogenations. The hydrogen atoms of primary and secondary amines are readily replaced by halogens, particularly in the presence of alkali. Some of the resultant ***N*-haloalkylamines** are quite unstable, decomposing, for example, to give nitriles (Scheme 2.36b).

$$\text{(a)}\quad Me_3N + H_2O_2 \longrightarrow Me_3N\rightarrow O + H_2O$$

$$\text{(b)}\quad RCH_2NH_2 \xrightarrow{Cl_2} RCH_2NCl_2 \xrightarrow{NaOH} RC\equiv N + 2NaCl$$

Scheme 2.36

Cleavage of the C–N Bond in Amines

Apart from the reactions of diazonium salts, a number of other reactions are known in which the C–N bond is broken. The best known of these is the **Hofmann elimination** of quaternary ammonium hydroxides (Scheme 2.37). An amine is converted by methylation with methyl iodide to the quaternary ammonium salt ('exhaustive methylation'). The iodide, on treatment with moist silver oxide, forms the quaternary ammonium hydroxide which undergoes a **bimolecular elimination** to form an alkene. The bimolecular elimination of 'onium' salts yields the least alkylated alkene. This substitution pattern is determined by the ease with which a hydrogen atom can be attacked by the base.

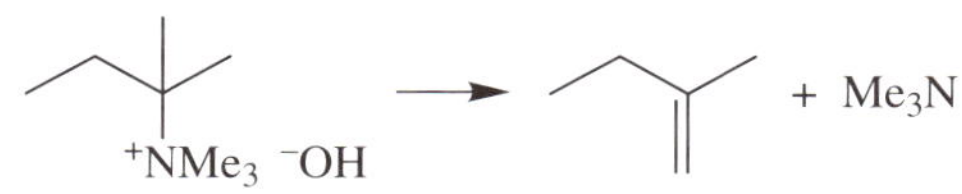

Scheme 2.37 Hofmann elimination

Exercise

Copy this revision chart showing the relationship between amines and other functional groups, and fill in the relevant reagents and conditions for the inter-relationships beside the arrows.

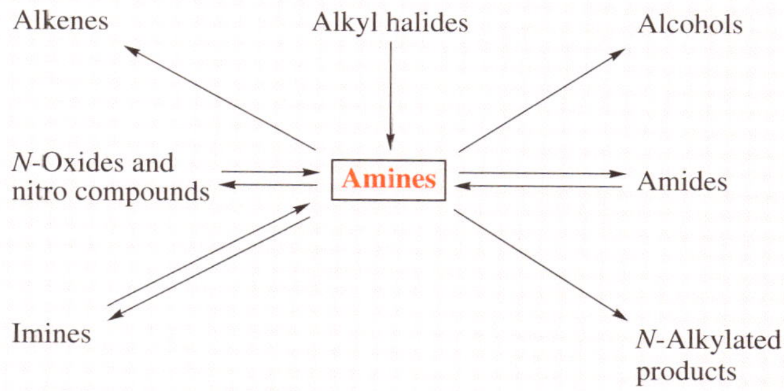

Summary of Key Points

1. Alkanes may be prepared by the hydrogenolysis of functional groups such as alkyl halides, by the reduction of unsaturated systems such as alkenes, alkynes and carbonyl compounds and by coupling reactions. The characteristic reactions of alkanes involve the abstraction of a hydrogen atom by a free radical to form a carbon radical.
2. Alkyl halides may be prepared from alcohols by substitution reactions and from alkenes by addition reactions.
3. Alkyl halides undergo nucleophilic substitution reactions by unimolecular (S_N1) and bimolecular (S_N2) pathways. The substitution reactions of alkyl halides, and of derivatives of alcohols, utilize oxygen, nitrogen, sulfur and carbon nucleophiles.
4. The elimination reactions of alkyl halides, alcohols and quaternary ammonium salts to form alkenes compete with substitution reactions.
5. Metal insertion reactions of alkyl halides produce organometallic derivatives such as Grignard reagents, which are synthetically useful sources of carbanions and change the reactive character of the carbon atom of the alkyl halide.
6. Alcohols may be prepared by the nucleophilic substitution of alkyl halides, the hydrolysis of esters, the reduction of carbonyl

compounds, the reaction of carbonyl compounds with organometallic reagents or the hydration of alkenes.

7. The hydroxyl hydrogen is weakly acidic. Alkoxides are strong bases and powerful nucleophiles. The oxygen lone pair can act as a base, accepting a proton or Lewis acid, and behaves as a nucleophile in the displacement of leaving groups and in addition reactions to carbonyl groups.
8. Alcohols are oxidized to aldehydes or ketones.
9. The reactivity of epoxides is dominated by the interaction of the lone pairs on the oxygen with electrophiles followed by the attack of a nucleophile on the carbon atom. The cleavage of epoxides takes place in a *trans* manner.
10. Thiols are more acidic than alcohols; the S–H bond also undergoes relatively easy homolytic fission. Thiols and thiolate anions are powerful nucleophiles.
11. Amines may be prepared by the nucleophilic substitution of alkyl halides with ammonia or other amines, by the reduction of imines, nitriles, amides, oximes or nitro compounds or by the rearrangement of amides.
12. The chemistry of amines is dominated by the nucleophilicity of the nitrogen atom. Amines form salts, participate in nucleophilic substitution reactions of alkyl halides and in nucleophilic addition reactions to carbonyl compounds.

Worked Problems

Q Explain why the rate of bimolecular substitution of 2,2-dimethylpropyl iodide, Me_3CCH_2I, by alkali is considerably slower than that of propyl iodide.

A The transition state for bimolecular substitution requires the hydroxide ion to approach from the rearside of the C–I bond (see **1**). The adjacent methyl groups hinder this approach.

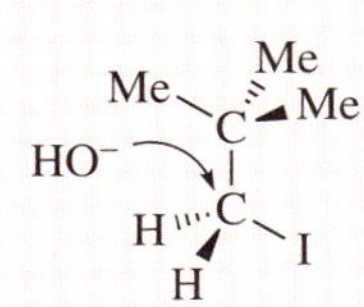

1

Q Explain why the treatment of 1-chloro-2-methylpropane with silver acetate gives 2-methylpropene and 2-acetoxy-2-methylpropane.

A The silver ion reacts rapidly with the chloride whilst the acetate anion is a relatively weak delocalized nucleophile. Consequently, the primary carbocation may rearrange to the more stable tertiary carbocation or eliminate a proton prior to attack by the nucleophile.

Q Explain why 2,2,2-trifluoroethanol is a stronger acid than ethanol.

A Fluorine is very electronegative and this strong electron-withdrawing effect not only polarizes the O–H bond in the sense O^{-}–H^{+} but also the electron-deficient carbon of the trifluoromethyl group stabilizes the alkoxide.

Q Treatment of pyrrolidine (**2**, C_4H_9N) with methyl iodide gave compound A, $C_6H_{14}NI$. Heating this with silver oxide gave compound B, $C_6H_{13}N$, which took up one mole of hydrogen on hydrogenation. What are the structures of A and B?

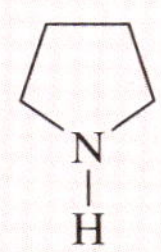

2

A Pyrrolidine is a secondary amine and reacts with two moles of methyl iodide to form a quaternary ammonium salt, compound A. The analytical data for compound B shows that it has one double bond equivalent which is present as an alkene. Consequently, compound B has been formed from compound A by a Hofmann elimination:

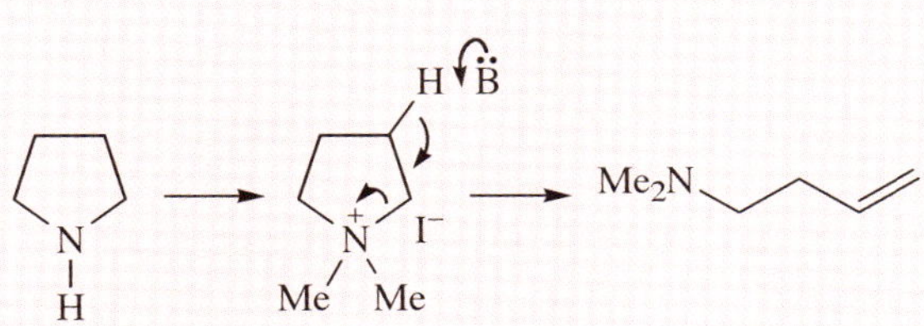

Problems

2.1 Give the products of the following reactions, all starting with Me_2CHCH_2I: (a) + KCN; (b) + AgCN; (c) + AgOAc; (d) + NaC≡CH; (e) + aq. KOH; (f) + alc. KOH; (g) + Ph_3P to give A and then NaH to give B; (h) + **3** to give C and H_2NNH_2 to give D; (i) + **4** to give E and then KOH to give F.

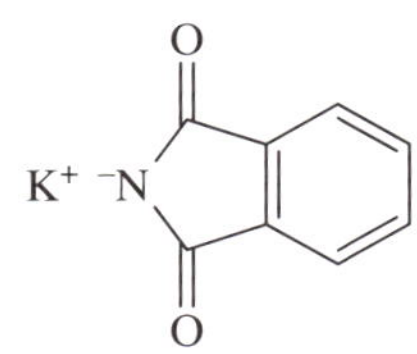

3

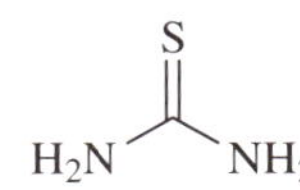

4

2.2 When compound A, C_3H_7Br, was heated with ethanolic potassium hydroxide, compound B, C_3H_6, was obtained together with some compound C, $C_6H_{14}O$. Treatment of compound B with hydrogen bromide in the dark gave compound D, which was an isomer of A. Vigorous oxidation of compound B with potassium permanganate gave compound E, $C_2H_4O_2$, and carbon dioxide. Identify the compounds A–E.

2.3 The following labelled alcohols were required for metabolic studies. Suggest ways in which they might be prepared starting from styrene ($PhCH=CH_2$): (a) $PhCH(OH)CH_2{}^2H$; (b) $PhCH^2HCH_2OH$; (c) $PhC^2H(OH)Me$; (d) $PhCH(^{18}OH)Me$.

2.4 Give the products of the following reactions and outline the mechanisms of the individual stages in the reactions.

(a) $Me(CH_2)_2OH + Na \rightarrow$ A and then EtI to give B

(b) $Me_2CHOH + MeCOCl \rightarrow$ C

(c) $Me_2CHOH + CrO_3 \rightarrow$ D

(d) Cyclohexanol ($C_6H_{11}OH$) + $Me_2SO + SO_3 + Et_3N \rightarrow$ E

(e) $Me(CH_2)_3OH + (PhO)_3PMe^+I^- \rightarrow$ F

2.5 Compound A, $C_4H_{10}O$, has infrared absorption at $\nu_{max} = 3550$ cm^{-1} and a two-proton triplet in its ^1H NMR spectrum at $\delta_H = 3.65$. On treatment with thionyl chloride, compound A gives B, C_4H_9Cl, which on refluxing with alcoholic potassium hydroxide gives, *inter alia*, C, C_4H_8. Compound C is oxidized by potassium permanganate to give D, $C_3H_6O_2$, which is soluble in sodium carbonate. Compound C reacts with HCl to give compound E, which is an isomer of B. Identify compounds A–E.

2.6 Explain the following observations: (a) hydrolysis of epoxycyclopentene (**5**) with dilute acid gave a mixture of 1,2-dihydroxycyclopent-3-ene (20%) and 1,4-dihydroxycyclopent-2-ene (80%); (b) treatment of the epoxide of 1,2-diphenylethene (stilbene) with boron trifluoride etherate gave an isomeric compound $C_{14}H_{12}O$, which formed a 2,4-dinitrophenylhydrazone and was oxidized to an acid, $C_{14}H_{12}O_2$.

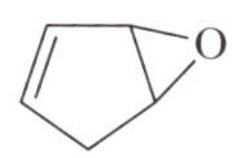

5

2.7 How might the following amines (a)–(d) be prepared free from contaminants ?

(a) 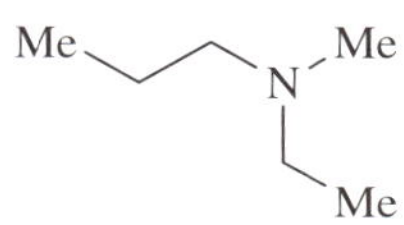

(b) 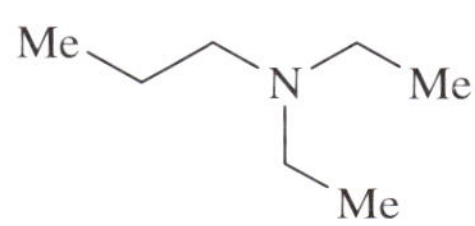

(c) 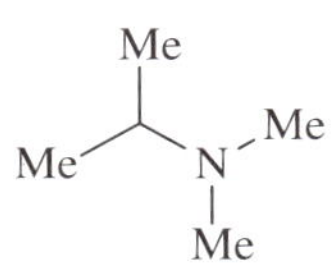

(d) 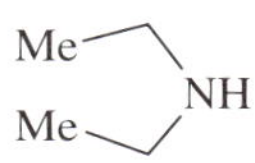

2.8 Why is not possible to prepare *t*-$BuNH_2$ and *t*-$BuCH_2NH_2$ by the action of NH_3 on the corresponding alkyl bromide? How may they be prepared from a related carboxylic acid?

2.9 Give the products of reactions (a)–(e):

(a) $MeNH_2 + ClCO_2Et \longrightarrow A$

(b) $EtNH_2 + (MeCO)_2O \longrightarrow B$

(c) 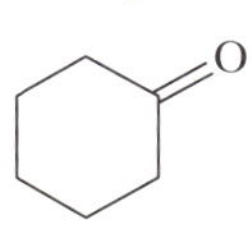 +  + $MeC_6H_4SO_3H \longrightarrow C$

(d) 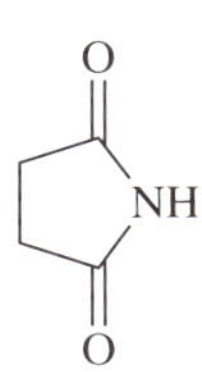 + KOBr + KOH $\longrightarrow$ D

(e) $MeNH_2 + 2ClCH_2CO_2H \longrightarrow E$

2.10 Arrange the nitrogen atoms (N_a, N_b and N_c) of histamine (**6**) in the order of their expected basicity.

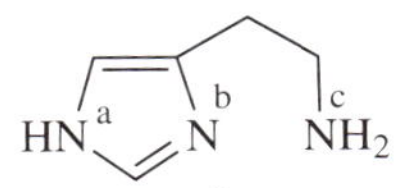

6

2.11 Place the compounds (a)–(d) in decreasing order of basicity.

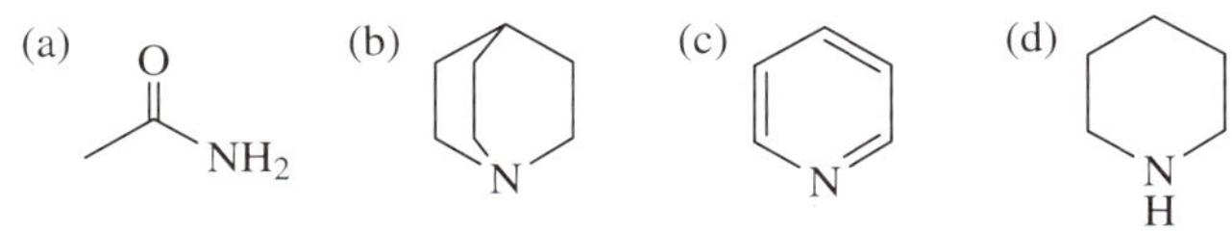

2.12 Suggest a mechanism for the following reaction:
$Me_2C(OH)CH_2NH_2 + HNO_2 \rightarrow MeC(O)CH_2Me$

3
Chemistry of the π-Bond

Aims

The aims of this chapter of the book are to describe the chemistry of the π-bond in various functional groups. By the end of this chapter you should be able to understand:

- The role of elimination and oxidation reactions in the formation of π-bonds
- The division of π-bonded functional groups into those with symmetrical and unsymmetrical π-bonds
- The electrophilic addition reactions of the electron-rich symmetrical π-bonds of alkenes
- The differences in reactivity between alkenes and alkynes
- The role of an electronegative oxygen or nitrogen atom in creating electron deficiency in the carbonyl and imino groups
- The nucleophilic addition reactions of the carbonyl group
- The role of acid catalysis in the chemistry of the carbonyl group
- The interaction between a carbonyl group and an adjacent group
- The effect of a carbonyl group in rendering an adjacent X–H hydrogen acidic
- The reactivity of carbanions with carbonyl groups in forming C–C bonds

The overlap between the trigonal sp^2 hybridized orbitals of the two carbon atoms of an **alkene** leads to the planar π-bonded system of the alkene and leaves an electron in a p orbital on each of the carbons. Overlap between these p orbitals leads to the π-system of the alkene. There is a region of increased electron density above and below the plane of the alkene. Hence the reactions of the alkene are typically those of electrophilic and radical addition.

Perturbation of the π-system by replacement of one carbon atom by the electronegative oxygen leads to an unsymmetrical distribution of the electrons between the atoms, and to electron deficiency on the remaining carbon atom. The **carbonyl group** (C=O) is thus sensitive to attack by nucleophiles on the carbon and to electrophilic attack on the lone pairs of the more electron-rich oxygen atom. If, however, an oxygen atom is attached to the alkene to give an **enol** (C=C–OH), the lone pairs on the oxygen atom may interact with the π-system of the alkene, thus increasing the electron-rich character of the double bond. The reactivity of the alkene towards electrophiles is enhanced on the carbon atom that is further from the oxygen atom. Similar effects are observed when nitrogen is introduced into the system.

The **tautomeric relationships** which exist between a ketone and an enol and between an imine and an enamine (see Chapter 1) are reflected in the susceptibility of the position α to a ketone to electrophilic attack. These effects are discussed in this chapter of the book.

3.1 Alkenes

3.1.1 Preparation of Alkenes

Alkenes may be obtained by elimination reactions from alkyl halides, alcohols, sulfonates or amines. The substitution pattern of the alkene and the stereospecificity of these methods depend quite subtly on the structure of the individual substrate. If the leaving group occupies an unsymmetrical position in a compound, one of two isomeric alkenes can be formed (Scheme 3.1). Elimination to give an alkene bearing the greatest number of alkyl groups is known as the **Saytzeff elimination**, and is commonly found with alkyl halides. When the elimination gives preferentially the less-substituted alkene, it is known as the **Hofmann elimination**. This substitution pattern is commonly found with the elimination of alkylammonium salts. In practice, mixtures of alkenes are often obtained.

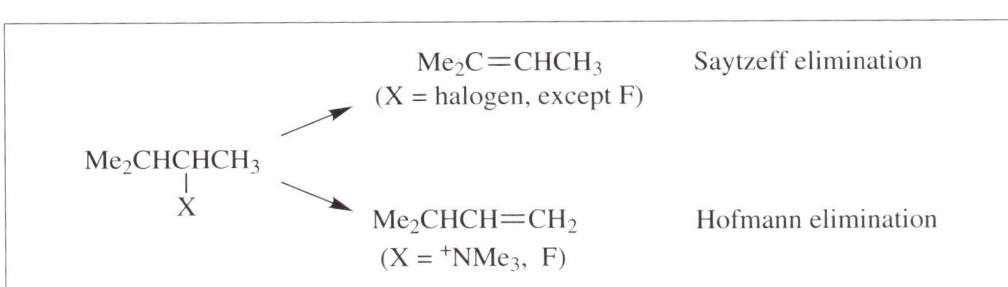

Scheme 3.1

A second general family of methods is based on the addition of carbanions to carbonyl groups. The particular alkene which is formed depends on the mechanism of the reaction. The **Wittig reaction** leads to the direct replacement of the carbonyl group by an alkene. The synthetic value of the reaction lies in the regiospecific formation of the alkene. Thus cyclohexanone gives only methylenecyclohexene on treatment with methylenetriphenylphosphorane. The alternative sequence, involving the addition of methylmagnesium iodide and dehydration of the resultant tertiary alcohol, gives a mixture of double bond isomers (Scheme 3.2).

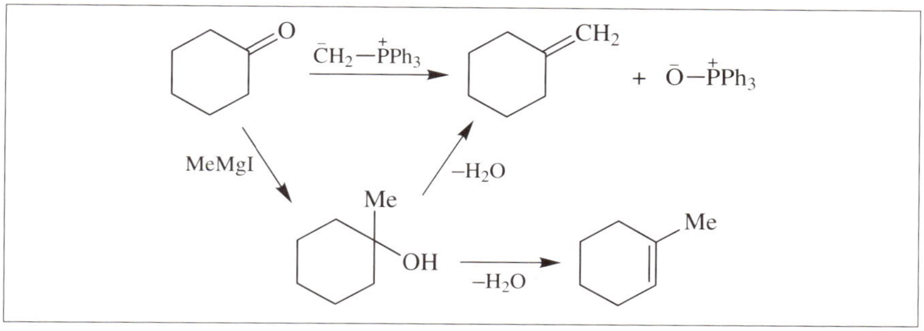

Scheme 3.2

However, although the Wittig reaction is regiospecific, *i.e.* the alkene is formed in a particular position in the molecule, it is stereoselective rather than stereospecific. Both *cis* and *trans* geometrical isomers can be formed. The ratio of geometrical isomers may be varied. The phosphonate variant of the process (the Wadsworth–Emmons reaction) leads to *trans* alkenes.

The partial reduction of alkynes provides methods that are both regio- and stereospecific. Dissolving metal reductions tend to give *trans* alkenes, whereas catalytic methods of reduction generate the *cis* alkenes (Scheme 3.3). A Lindlar catalyst ($Pd/CaCO_3$ + PbO, partially poisoned with quinoline) has been recommended for use in this context.

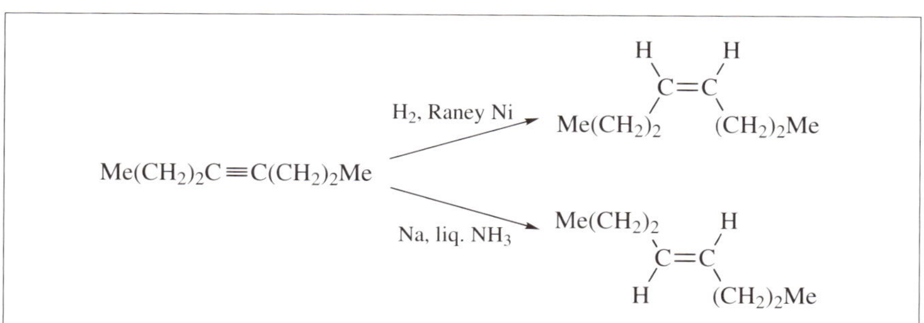

Scheme 3.3

3.1.2 Reactions of Alkenes

The reactions of alkenes are determined by the **electron-rich π-system** of the alkene. Mechanistically, these reactions may be divided into a number of groups. The π-system may react with electrophiles, with radicals and as a component in pericyclic reactions, including acting as a dienophile in Diels–Alder reactions. The alkene may also stabilize an adjacent electron-deficient carbon either as a radical or as a carbocation. This pattern of reactivity is summarized in Box 3.1.

Box 3.1 Reactivity of Alkenes

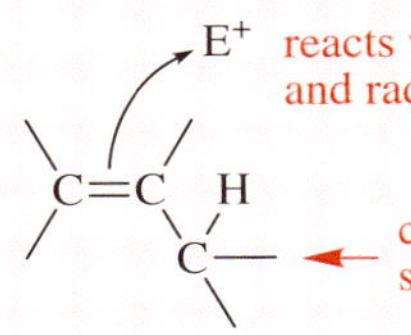

Electrophilic Addition

The reactions which involve **electrophilic addition**, followed by reaction with the associated nucleophile, commonly result in *trans* addition arising from a requirement for the reacting orbitals to lie in a plane. A second electronic feature determines the regiochemistry of electrophilic additions. The electron-releasing effect of an alkyl group stabilizes an adjacent carbocation. Hence, in those electrophilic additions in which there is some intermediate carbocationic character, the addition of the electrophile takes place to generate the more highly substituted carbocationic intermediate. This provides a mechanistic interpretation of the **Markownikoff rule**, which states that "in the addition of HX to an alkene, the hydrogen atom (the electrophile) becomes attached to the less-substituted carbon atom".

The addition of radicals follows a different pattern which is determined by access to the alkene. The addition of hydrogen bromide under photochemical conditions, or in the presence of peroxides, involves the initial attack of a bromine radical on the alkene. The consequences are illustrated by the reaction of hydrogen bromide with allyl bromide (Scheme 3.4).

Another family of addition reactions to alkenes involves a cyclic intermediate or a cyclic transition state. These commonly lead to *cis*

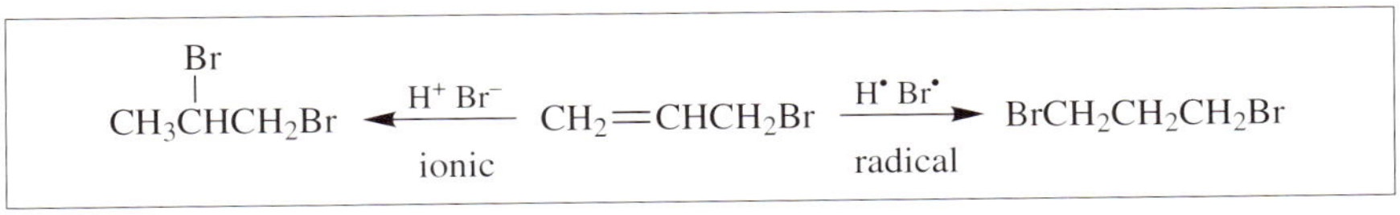

Scheme 3.4

addition. This series also includes the cycloaddition of dienes and 1,3-dipolar compounds.

The **facial selectivity** of the addition to an alkene may be determined by a number of factors. The face of the alkene which is attacked by an electrophile may be determined by the steric bulk of neighbouring groups. A group such as an alkyl group may protect one face, so directing the incoming reagent to the opposite face. Secondly, if a bulky cyclic intermediate is formed during the reaction, the face on which this occurs may be determined by steric interactions. On the other hand, hydrogen bonding between a neighbouring hydroxyl group in the substrate and the reagent may lead to the initial addition taking place on the same face as the directing group, in this case the hydroxyl group.

The electron density on the two faces of the alkene may be selectively modified by adjacent substitution. For example, the repulsion between the lone pairs of an adjacent ether and the π-system may increase the electron density on the face opposite to the ether. Hyperconjugative assistance from adjacent axial C–H bonds may favour attack from a particular face of an alkene.

The classical examples of electrophilic addition are those of the addition of the hydrogen halides and are exemplified by the addition of hydrogen chloride under ionic conditions, when the products are those of Markownikoff addition (Scheme 3.5).

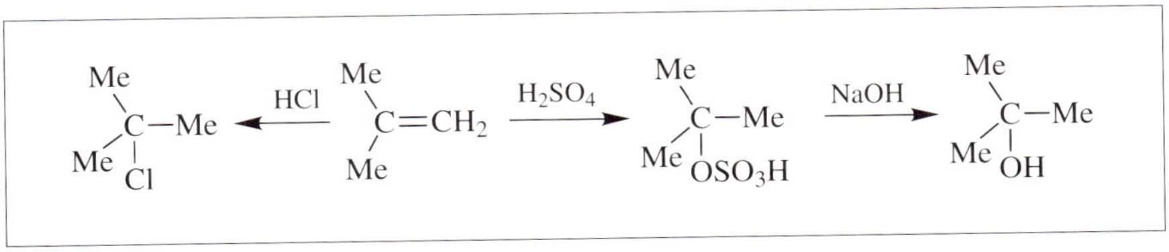

Scheme 3.5

The initial addition of the electrophilic proton is not always followed by the addition of a nucleophile. The carbocationic intermediate may be discharged by the loss of a proton from another position, leading to **double bond isomerization** (Scheme 3.6).

Secondly, the carbocation may act as an electrophile and react with

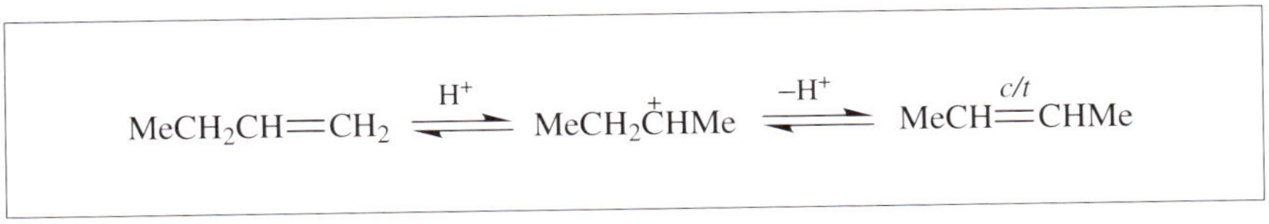

Scheme 3.6

another electron-rich centre within the molecule, for example another double bond or an aromatic ring. This can lead to the formation of new C–C bonds and, in appropriate circumstances, to **cyclization reactions** (Scheme 3.7).

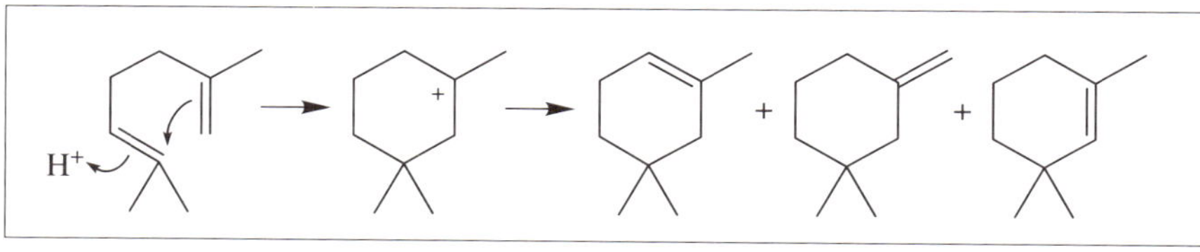

Scheme 3.7

Thirdly, the carbocation may initiate a series of **rearrangement reactions** with either the eventual loss of a proton or attack by a nucleophile elsewhere in the molecule (Scheme 3.8). The classical examples of the Wagner–Meerwein rearrangements of terpenoids follow this pattern.

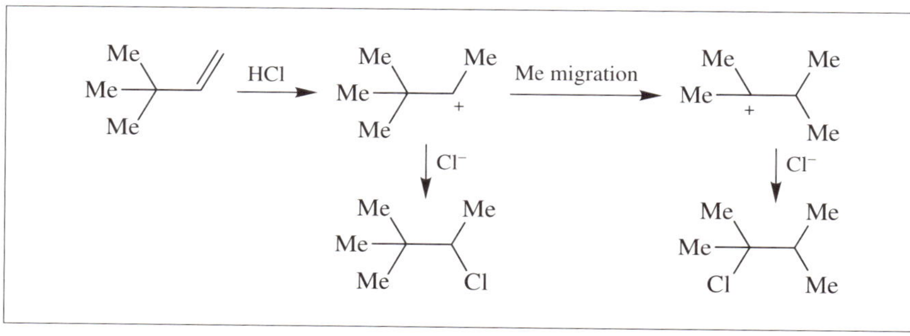

Scheme 3.8

Halogen Electrophiles

The additions of halogens such as bromine to alkenes represent a good test for unsaturation in a molecule. The addition of bromine takes place via the formation of a **bromonium ion**. Proof of this was obtained by carrying out the reaction of ethene with bromine in the presence of sodium chloride, when the product was the bromo-chloride ($BrCH_2CH_2Cl$).

The addition of bromine takes place in a *trans* manner. In a cyclic system this may lead to **diaxial dibromides** (Scheme 3.9).

The incoming nucleophile may not necessarily be a second halogen atom. Hypobromous acid (HOBr), generated from *N*-bromosuccinimide

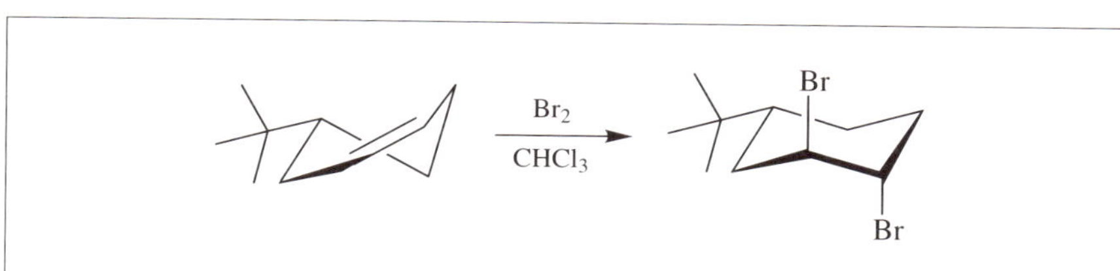

Scheme 3.9

and perchloric acid ($HClO_4$), behaves as a source of a bromonium ion, with the hydroxide ion reacting in the subsequent step (Scheme 3.10).

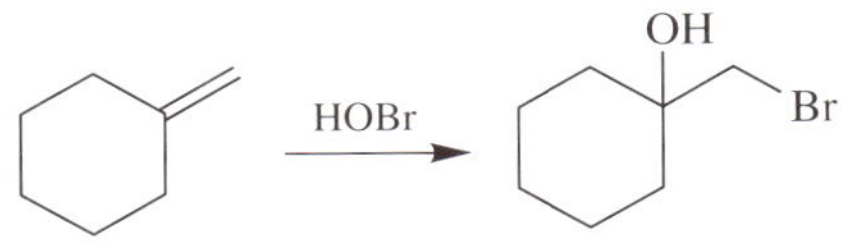

Scheme 3.10

Nitrogen Electrophiles

The **nitronium ion** (NO_2^+) is a powerful electrophile and can nitrate alkenes. Although the reaction may be followed by the attack of a nucleophilic component, it is often completed by the loss of a proton and the formation of a **nitroalkene** (Scheme 3.11).

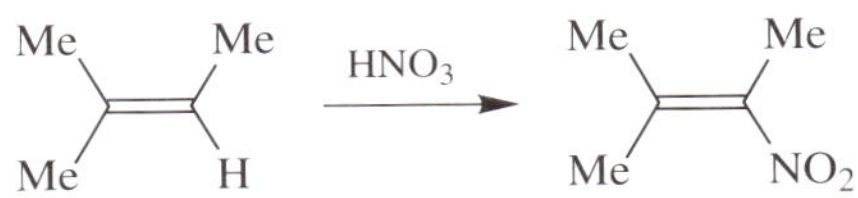

Scheme 3.11

The **nitrosonium ion** (NO^+) is another electrophile which is generated either from the oxides of nitrogen or from nitrosyl chloride. The nitrosochloride may **tautomerize** to chloro-oxime adducts.

Carbon Electrophiles

The reactions of alkenes with carbon electrophiles have already been mentioned in the cyclization of 1,5-dienes. However, carbon electrophiles may be generated in other ways. Protonation of formaldehyde (methanal) leads to a carbocation that may be stabilized by the oxygen lone pair (Scheme 3.12a). This may react with alkenes with the formation of 1,3-glycols or unsaturated alcohols, depending upon the way in which the intermediate carbocation is discharged (the **Prins reaction**, Scheme 3.12b).

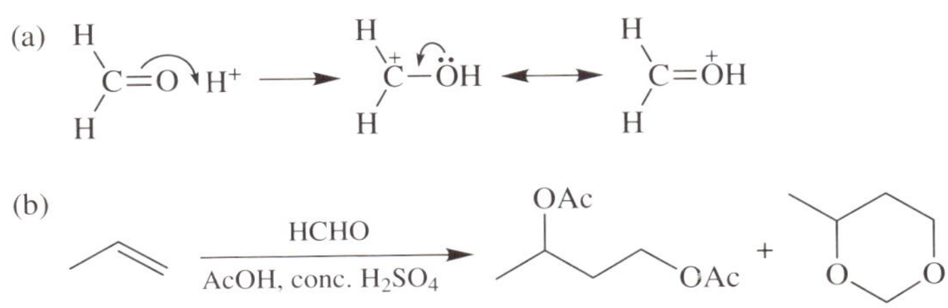

Scheme 3.12

The **Friedel–Crafts acylation** of an alkene is a useful reaction which can lead to the formation of an unsaturated ketone (Scheme 3.13).

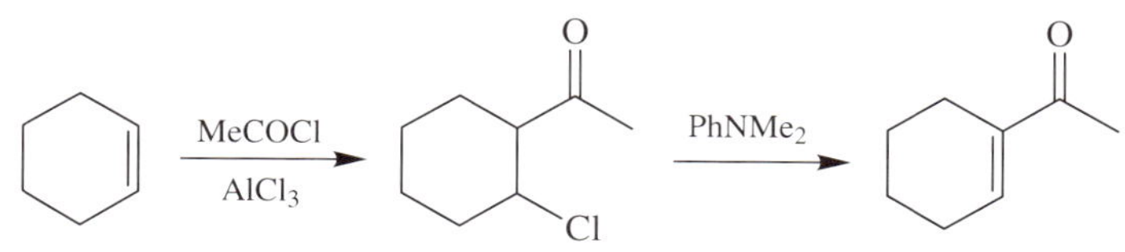

Scheme 3.13

Oxymercuration

A number of metals salts can be used as the source of electrophiles in reactions with alkenes. One of the most interesting of these involves the attack of mercury(II) acetate in acetic acid. Reductive cleavage of the organomercury compound with sodium borohydride leads to the overall **hydration** of the alkene in a Markownikoff sense. There are a number of preparative advantages, such as a reduced tendency to rearrange, associated with this and similar relatively mild procedures when compared to the direct protonation of a double bond (Scheme 3.14)

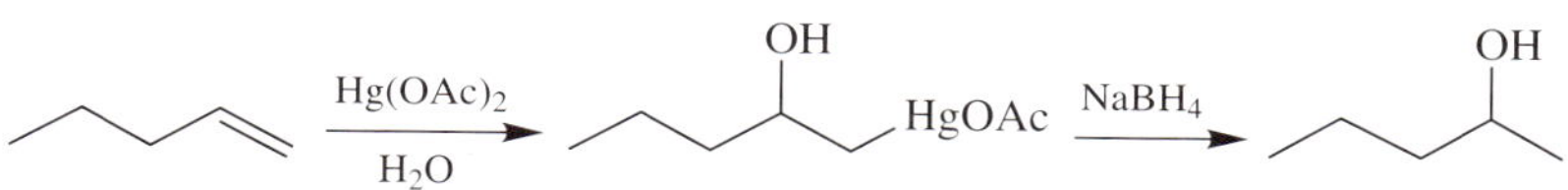

Scheme 3.14

Reactions involving *syn* Addition

The **catalytic hydrogenation** of a double bond involves the adsorption of the alkene on a metal surface and the transfer of hydrogen from the surface to the double bond. Typical catalysts are finely divided forms of nickel, platinum or palladium, the latter often supported on an inert carrier such as charcoal or barium sulfate. Hydrogenations are carried out in solution, with the hydrogen at atmospheric or higher pressure. The addition of hydrogen is typically *cis* and from the less-hindered face of the molecule (*e.g.* the hydrogenation of α-pinene, **3.1**).

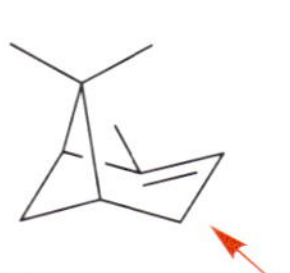

3.1 α-pinene

The ability of alkenes such as ethene to form stable metal complexes such as **Zeise's salt** ($K[PtCl_3(C_2H_4)]$) has been exploited in the development of soluble metal catalysts for reduction. The catalyst is a soluble metal complex in which the central metal atom has the ability to bind both hydrogen and an alkene. The best known of these catalysts is **Wilkinson's catalyst** [$RhCl(PPh_3)_3$].

The *cis* addition of borane to an alkene and the subsequent transformations of the **alkylboranes** are reactions of major synthetic importance.

Although the reagent is often the borane–tetrahydrofuran complex (BH_3:THF), other more highly substituted, more sterically demanding, alkylboranes (RBH_2 and R_2BH) are also used to enhance the stereoselectivity of the reaction.

The electrophile is the electron-deficient boron. The reaction takes place in several stages. The first stage involves the formation of a π-complex between the alkene and the borane. The stereochemical directing effects of neighbouring alkyl groups and lone pairs on adjacent oxygen atoms determine the facial selectivity at this stage. The second stage involves the rearrangement of the π-complex to form an alkylborane. The regiochemistry of this is determined by the stabilization of carbocations at the respective carbon atoms. The electron-deficient boron atom behaves in the same way as a proton, resulting in Markownikoff addition of borane across the double bond.

The synthetic value of the reaction lies in the modification of these organoboranes. The commonest reaction involves the decomposition of the borane by alkaline hydrogen peroxide. The highly nucleophilic hydroperoxide anion attacks the electron-deficient boron with the formation of an 'ate' complex. Rearrangement of this leads to the formation of a borate ester which then undergoes hydrolysis to an alcohol in which an oxygen atom has replaced the boron (Scheme 3.15). The overall outcome of this reaction is the **anti-Markownikoff hydration** of the double bond. The regiochemistry is the reverse of the acid-catalysed hydration of an alkene. The overall addition of water takes place in a *cis* manner on the less-hindered face of the double bond.

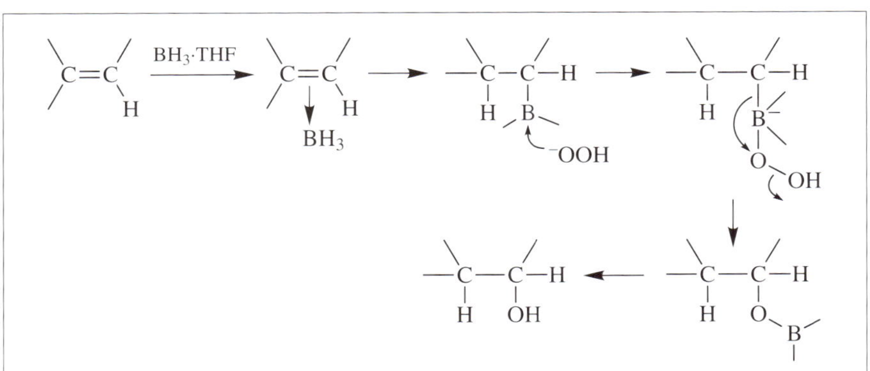

Scheme 3.15

Other useful reactions of the boranes include **protonolysis** of the borane with a carboxylic acid, which leads to hydrogenation of the alkene.

The reaction of alkenes with peroxy acids involves an initial complex being formed between the alkene and the oxygen of the peroxy acid. Further collapse of the intermediate leads to the formation of an

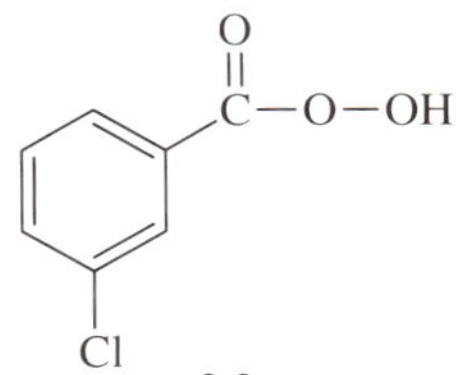

3.2

epoxide (oxirane) (Scheme 3.16). The peroxy acid that is used may be *m*-chloroperbenzoic acid (**3.2**) or peracetic acid. Alternatively, the peroxy acid may be generated *in situ* from hydrogen peroxide and an acid anhydride.

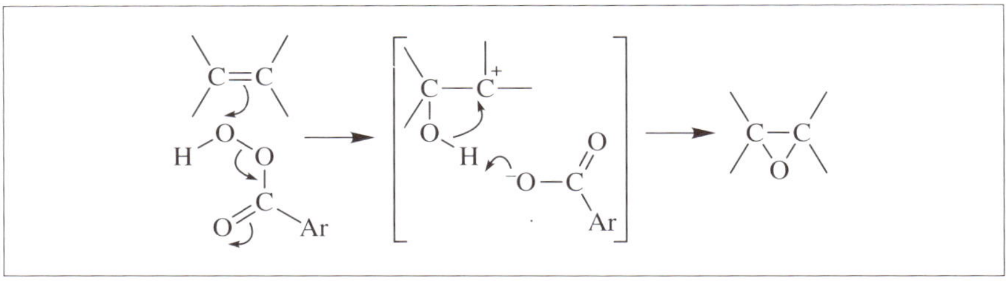

Scheme 3.16

The facial selectivity of epoxidation is dominated by steric factors. Steric hindrance by neighbouring alkyl groups may direct attack to the opposite face, while hydrogen bonding between the reagent and an adjacent hydroxyl group may bring about the addition of the oxygen to same face of the alkene as the neighbouring hydroxyl group.

Epoxides may also be formed from alkenes by the treatment of halohydrins with base.

Whereas hydrolysis of epoxides leads to the *trans* diaxial addition of water and the formation of *trans* glycols (1,2-diols), *cis* glycol formation involves the addition of osmium(VIII) oxide (osmium tetroxide, OsO_4) or cold dilute aqueous potassium manganate(VII) (potassium permanganate) to an alkene.

A number of mechanisms have been proposed for osmylation involving the formation of the osmate ester. Subsequent hydrolysis of the ester gives the *cis* glycol (Scheme 3.17). The process can be converted into a catalytic sequence by the re-oxidation of the osmium using a variety of reagents such as hydrogen peroxide, *t*-butyl hydroperoxide, morpholine *N*-oxide or potassium hexacyanoferrate(III) (potassium ferricyanide). Although alkaline potassium permanganate can be used as a much cheaper alternative, the reaction is often more complex. These *cis* glycols are readily cleaved by sodium iodate(VII) (sodium periodate).

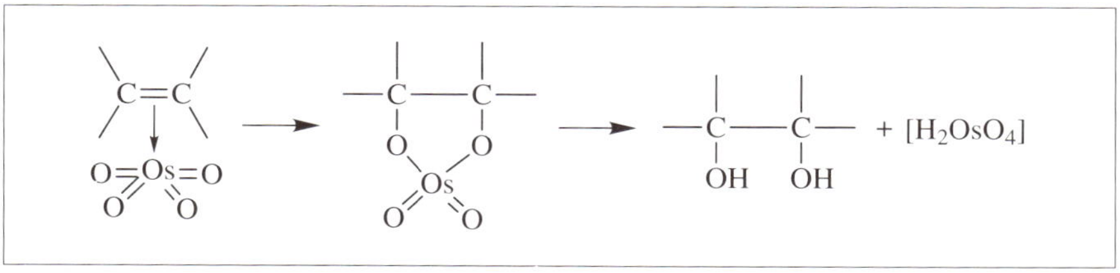

Scheme 3.17

Pericyclic Reactions

There are a number of compounds whose structures can be represented by a combination of resonance forms to which a 1,3-dipole (a^+–b–c^-) makes a significant contribution. These dipolar compounds participate in cycloaddition reactions with alkenes. A common 1,3-dipolar cycloaddition is the reaction with ozone (O_3), which rapidly adds to an alkene to form a molozonide. However, the initial adduct is unstable because of the repulsion between the oxygen lone pairs and the resultant weakness of the O–O bond. Rearrangement takes place with the formation of the ozonide. Cleavage of the ozonide by various reagents such as zinc dust or triphenylphosphine may lead, depending on the vigour of the reductant, either to dicarbonyl compounds or to alcohols (Scheme 3.18).

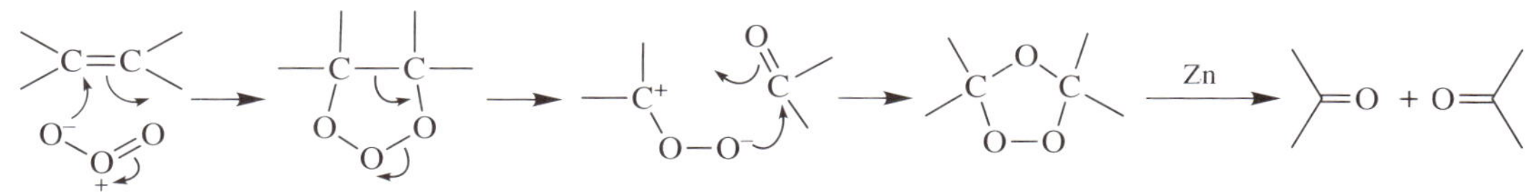

Scheme 3.18

Many other 1,3-dipolar cycloadditions are known, amongst which is the addition of diazomethane (CH_2N_2). Expulsion of nitrogen from the adduct leads to the formation of a cyclopropane ring. Another way of achieving the same result involves the addition of a carbene such as the Simmons–Smith reagent. This is generated from methylene iodide (CH_2I_2) and a zinc/copper couple.

Diels–Alder and Related Reactions

The Diels–Alder reaction between a diene and a dienophile is one of the major synthetic reactions that are used for the formation of C–C bonds. Its usefulness lies in the mild conditions and the predictable regio- and stereospecificity of the process. For most purposes the thermal reaction requires the presence of an electron-withdrawing substituent on the alkene. The reaction is exemplified by the addition of the alkene of maleic anhydride as the dienophile to the diene of cyclopentadiene (Scheme 3.19).

A consequence of a Diels–Alder reaction is the formation of a cyclohexene. The reverse of this reaction, involving the fragmentation of a

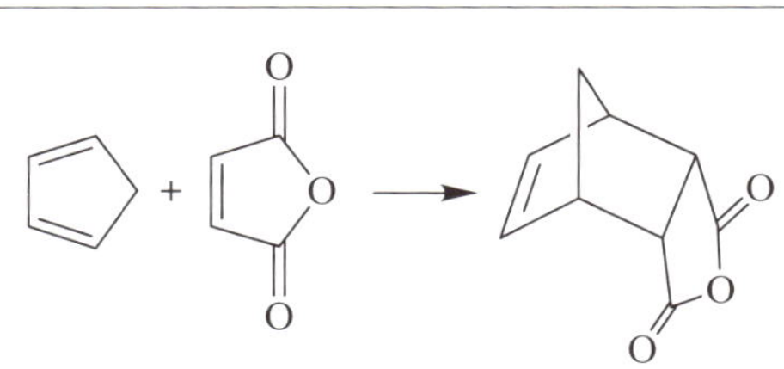

Scheme 3.19

cyclohexene under thermal conditions, may sometimes be observed. Cyclopentadiene is obtained by the thermal decomposition of its dimer in a retro-Diels–Alder reaction.

The **ene reaction** is another pericyclic reaction which involves the combination of an alkene with an allylic system and a hydrogen atom transfer (Scheme 3.20).

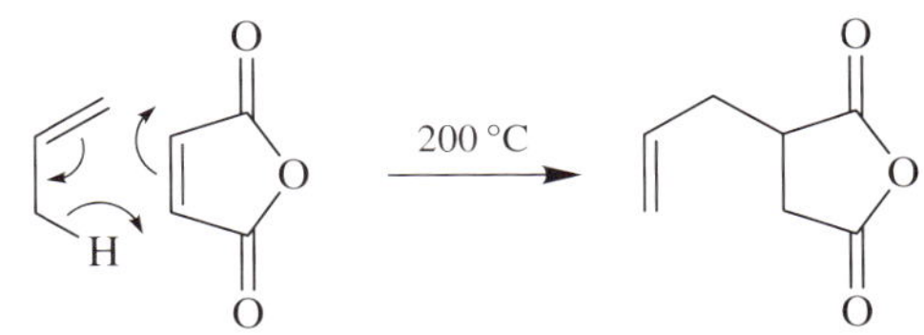

Scheme 3.20

Polymerization Reactions

$CH_2{=}CH_2$ ethene

$-(CH_2-CH_2)_n-$ polyethene **3.3**

$MeCH{=}CH_2$ propene

$-(CH(Me)-CH_2)_n-$ polypropene **3.4**

The **polymerization** of alkenes is of major economic importance. It may take place with either **free radical** or **ionic catalysis**. Three stages may be distinguished in the polymerization process: initiation, propagation and termination. The free radical polymerization of ethene under pressure and at temperatures of 50–150 °C leads to low-density polyethene (**3.3**). The polymer is of relatively low density due to chain branching and the consequent poor packing of the chains. Typical initiators are peroxides such as dibenzoyl peroxide, *t*-butyl hydroperoxide, oxygen or azobisisobutyronitrile. A more regular polymer known as high-density polyethene is formed when catalysts based on titanium(IV) chloride and triethylaluminium (the Ziegler catalysts) are used. These catalysts may be used with propene to give polypropene (**3.4**).

Alkene Metathesis

Alkene metathesis is a reaction catalysed by transition metals in which the carbon atoms that constitute the double bond of the alkene are exchanged with those of another alkene *via* a metal–carbene complex (Scheme 3.21).

$$2RCH{=}CHR' \xrightarrow{\text{cat.}} RCH{=}CHR + R'CH{=}CHR'$$

Scheme 3.21

The combination of short- and long-chain alkenes to make alkenes of medium chain length (C_{11}–C_{14}) has been important in the petrochemical industry. Typical catalysts are based on complexes of molybdenum, tungsten or ruthenium (**3.5**). The application of the reaction in alkene synthesis often arises from the formation of volatile, easily separable alkenes, such as ethene, as the second product (Scheme 3.22). Cross metathesis between two acyclic alkenes can also be useful, particularly in ring-closing alkene metathesis.

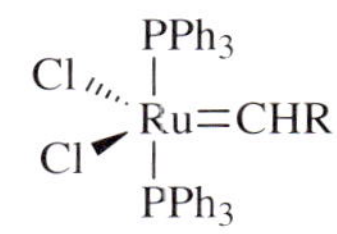

3.5

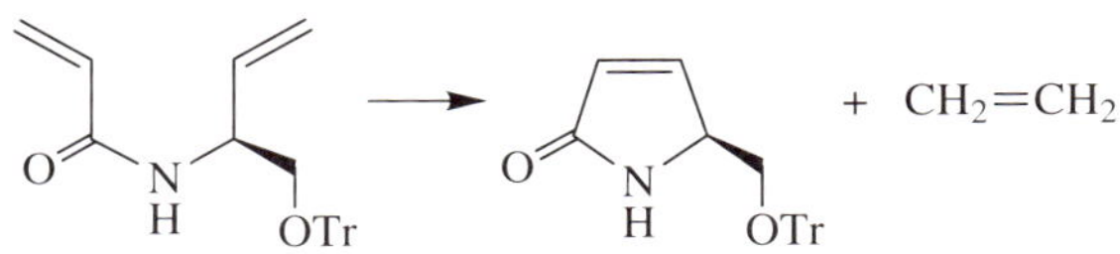

(Tr = triphenylmethyl protecting group)

Scheme 3.22

Reactivity at the Allylic Position

An alkene may enhance the reactivity of an adjacent allylic position by making a hydrogen atom, or a substituent, labile and by stabilizing an adjacent radical or carbocation (Scheme 3.23).

$$RCH{=}CH{-}\overset{+}{C}H_2 \longleftrightarrow R\overset{+}{C}H{-}CH{=}CH_2$$

Scheme 3.23

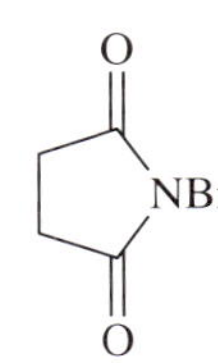

3.6

Bromination with a free-radical brominating agent such as *N*-bromosuccinimide (**3.6**) takes place at the allylic position. The radical that is formed in the first step of the reaction is delocalized, and hence the final product may be a mixture (Scheme 3.24).

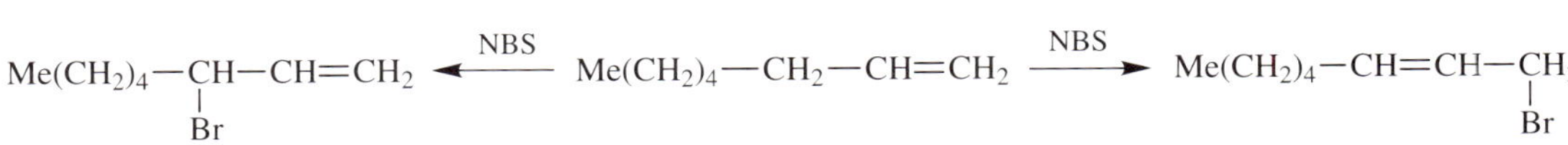

Scheme 3.24

Oxidation with chromium(VI) oxide, *t*-butyl chromate or chromyl chloride may also take place at the allylic position with the formation of unsaturated ketones. The alkene may enhance the rate of displacement of an allylic halide, and the reaction may even take place with the participation of the double bond, so leading to an allylic rearrangement (Scheme 3.25).

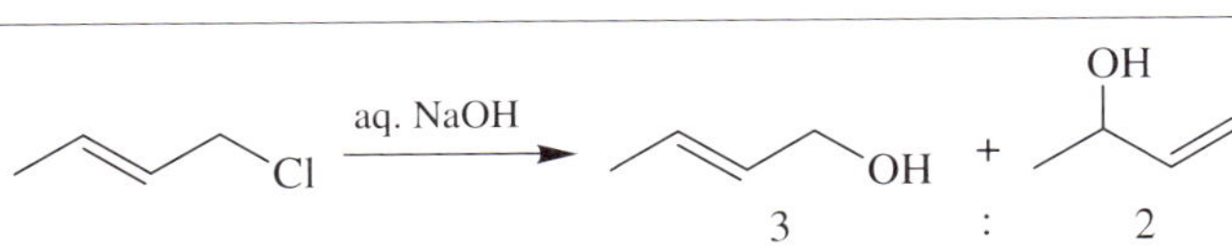

Scheme 3.25

Exercise

Copy this revision chart showing the relationship between alkenes and other functional groups, and fill in the relevant reagents and conditions for the inter-relationships beside the arrows.

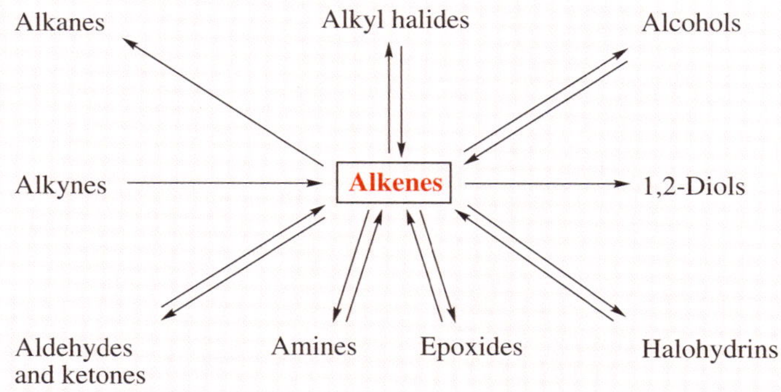

3.2 Alkynes

3.2.1 Preparation of Alkynes

Alkynes (acetylenes, RC≡CR) may be prepared by the elimination of a hydrogen halide from alkenyl halides under vigorous conditions. This is exemplified by the preparation of phenylacetylene from cinnamic acid *via* the dibromide and ω-bromostyrene (Scheme 3.26). The contrast between the conditions required for the bromodecarboxylation and for the second elimination to form the alkyne reveals the difference in reactivity between an alkyl and an alkenyl halide. Alternative modes of elimination, such as allene formation or rearrangement reactions, restrict the use of this procedure.

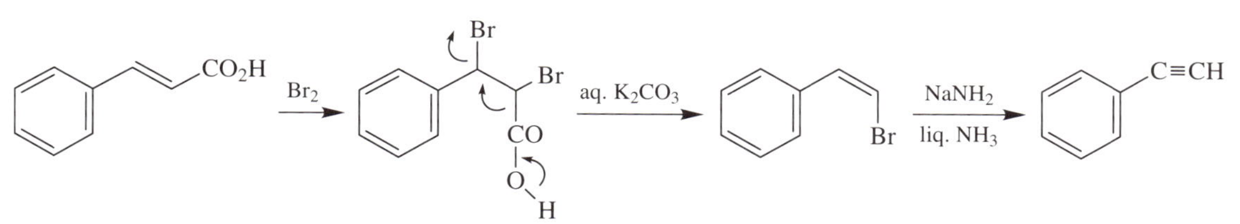

Scheme 3.26

A common method for the preparation of alkynes is to insert acetylene (ethyne) itself into a molecule *via* the acetylide (Scheme 3.27a) or Grignard reagent (Scheme 3.27b).

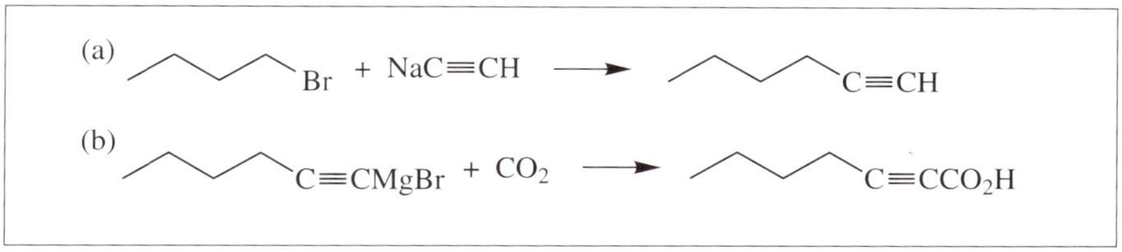

Scheme 3.27

3.2.2 Addition Reactions of Alkynes

Alkynes are highly unsaturated and are reactive towards electrophiles and nucleophiles. They are less reactive than alkenes towards electrophiles and more reactive than alkenes towards nucleophiles.

The reduction of alkynes may be carried out with sodium dissolving in liquid ammonia. Subsequent protonation of the dianion gives the *trans*-alkene. On the other hand, catalytic reduction gives the *cis*-alkene. An illustration of alkyne chemistry involves the preparation of intermediates for the synthesis of vitamin A shown in Scheme 3.28. Complete hydrogenation to an alkane takes place over a platinum catalyst.

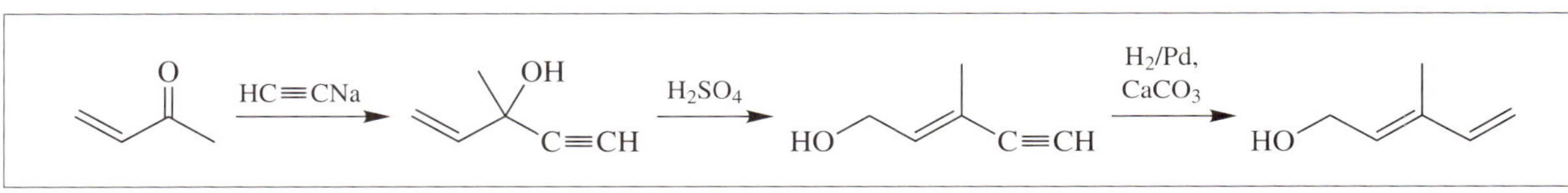

Scheme 3.28

In the absence of a catalyst, alkynes react very slowly with bromine, particularly when compared to alkenes. When a choice exists, bromine reacts preferentially with an alkene rather than an alkyne. It is possible that radical reactions play a more important role in the addition to alkynes. When the reaction of acetylene with chlorine is catalysed by iron(III) chloride, the reaction is fast and 1,1,2,2-tetrachloroethane is formed. The uncatalysed addition of a hydrogen halide gives a *trans* alkenyl halide. Further addition is restricted but can give rise to dihalides.

The mercury-catalysed hydration of alkynes is a useful reaction which leads, *via* protonolysis of the organomercury compound and the enol, to a methylene ketone (Scheme 3.29).

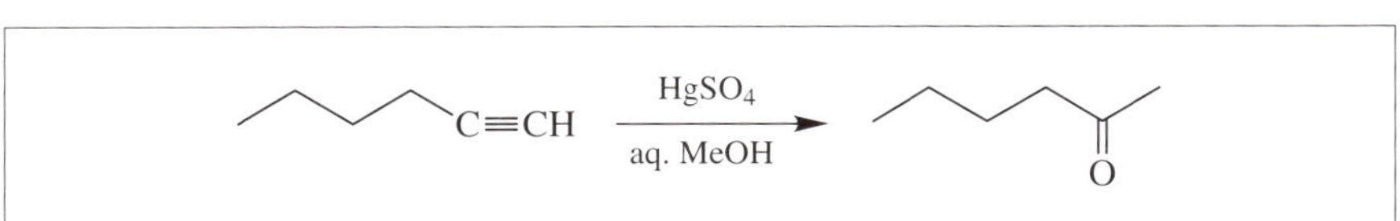

Scheme 3.29

Some alkynes undergo **nucleophilic addition**. Thus acetylene adds methanol under pressure in the presence of sodium methoxide to form methyl vinyl ether (methoxyethene).

3.2.3 Formation of Acetylides

The increase in 's' character of the C–H bond (sp), compared to that of an alkene (sp^2) or an alkane (sp^3), means that the electrons of the C–H bond are held more tightly by the carbon, thus making the hydrogen atom more acidic. In the presence of a strong base, typically sodamide in liquid ammonia, acetylene may form an acetylide. The acetylide is a powerful nucleophile and will add to carbonyl groups. These reactions are useful C–C bond-forming processes (see Scheme 3.28).

Some of the organometallic acetylides, particularly the copper derivatives, can be linked together by oxidation to form diacetylenes in the Cadiot–Chodkiewick oxidative coupling reaction (Scheme 3.30).

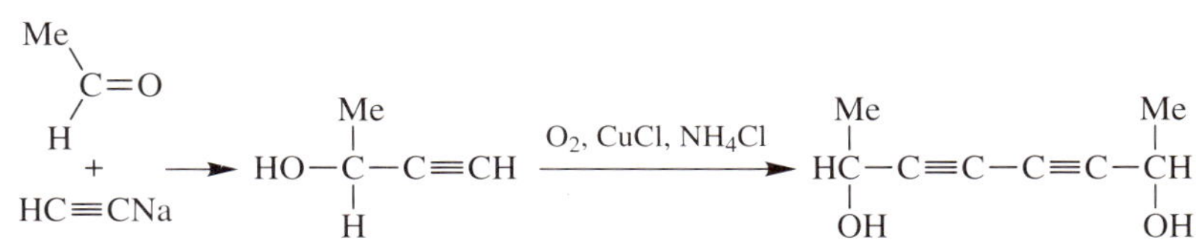

Scheme 3.30

3.2.4 Metal Complexes of Alkynes

A number of metal carbonyls and cyanides, particularly those of nickel and iron, form π-complexes with alkynes. These systems behave catalytically in the carbonylation of acetylene and in the formation of trimers (benzene) and tetramers (cyclooctatetraene).

3.2.5 Rearrangements of Alkynes

Alkynes may participate in a number of rearrangement reactions, exemplified by the Rupe rearrangement of ethynylcyclohexanol to an unsaturated ketone (Scheme 3.31).

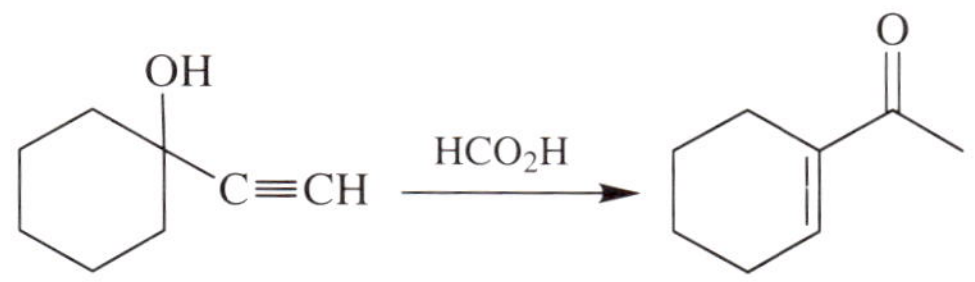

Scheme 3.31

Anionotropic rearrangements and displacement reactions yield allenes. Allenes are also formed in basic media (Scheme 3.32). The reaction is particularly easy if the migrating hydrogen is adjacent to a second multiply bonded carbon.

$$PhC{\equiv}CCH_2Ph \xrightarrow[20\,°C,\ 20\ min]{Al_2O_3,\ KOH} PhCH{=}C{=}CHPh$$

Scheme 3.32

Exercise

Copy this revision chart showing the relationship between alkynes and other functional groups, and fill in the relevant reagents and conditions for the inter-relationships beside the arrows. Note the geometry of the alkenes that are formed.

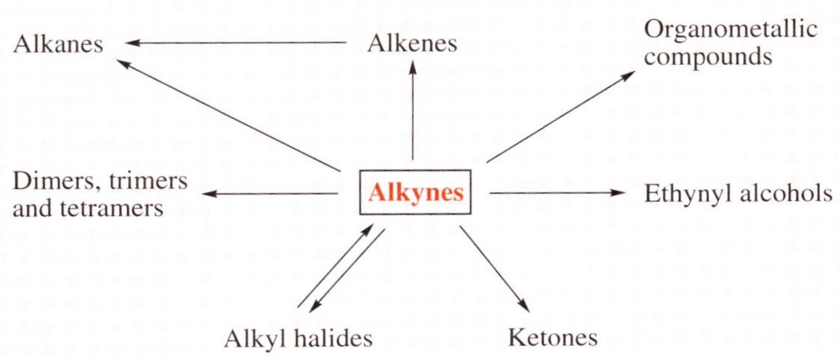

3.3 Carbonyl Compounds

Dipole moment measurements of carbonyl compounds have shown that there is an uneven distribution of charge within the carbonyl group, with the electronegative oxygen bearing more negative charge and the carbon atom being electron deficient. Electron donation to the electron-deficient carbon atom may be provided by an external nucleophile (Scheme 3.33a), by an adjacent lone pair (Scheme 3.33b) or by an adjacent anion (Scheme 3.33c).

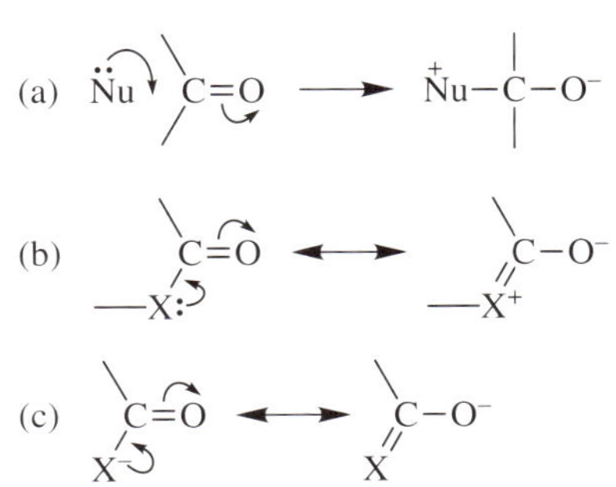

Scheme 3.33

The oxygen atom is basic and reacts with electrophiles, such as the proton and Lewis acids. Protonation of the oxygen increases the electron deficiency of the carbon atom of the carbonyl group. Hence many addition reactions of the carbonyl group are acid catalysed (Scheme 3.34).

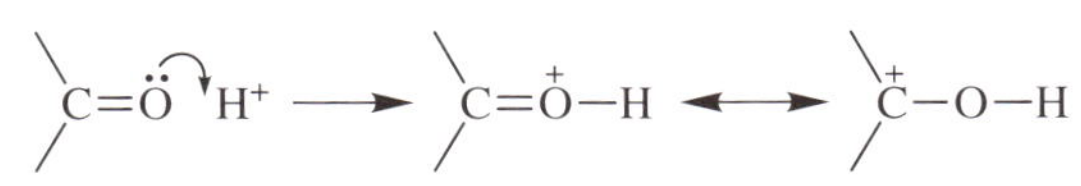

Scheme 3.34

The carbonyl group has an influence on the chemistry of substituents. The electron donation from the lone pairs on oxygen and nitrogen in esters and amides not only diminishes the reactivity of the carbonyl group towards nucleophiles, but also reduces the basicity of the oxygen and nitrogen atoms.

The electron-withdrawing effect of the carbonyl group makes hydrogen atoms attached to neighbouring atoms acidic (Box 3.2). Once the anion is formed, it may achieve stability by delocalization of the negative charge over the carbonyl group.

Box 3.2 Acidic Hydrogens

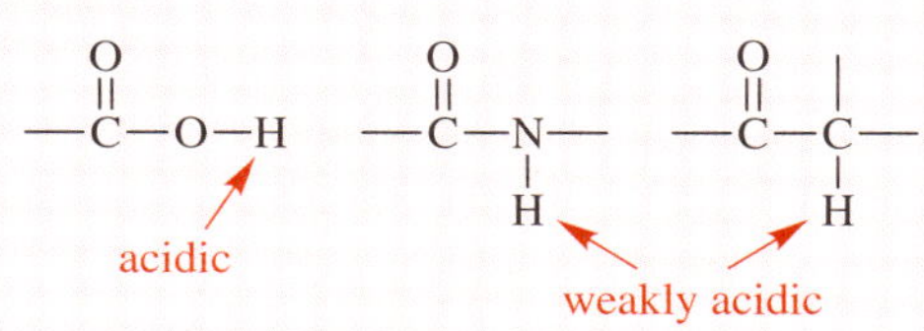

There is a **tautomeric relationship** between the carbonyl compound and the corresponding enol. The latter possesses an electron-rich alkene. The hydrogen atom of the enolic hydroxyl group is acidic and in the presence of base forms an anion (Scheme 3.35a). Protonation of the enolate anion may take place on oxygen or carbon to regenerate either the enol or the corresponding carbonyl compound. The existence of the electron-rich enolic form of carbonyl compounds leads to the position adjacent to a carbonyl group being sensitive to electrophilic attack (Scheme 3.35b).

The electron-withdrawing effect of the carbonyl group may be relayed through a conjugated system. Hence the β-carbon of an **α,β-unsaturated ketone** is susceptible to nucleophilic attack (Scheme 3.36). A hydroxyl group attached to this β-carbon is acidic and the anion is stabilized by delocalization (Scheme 3.37).

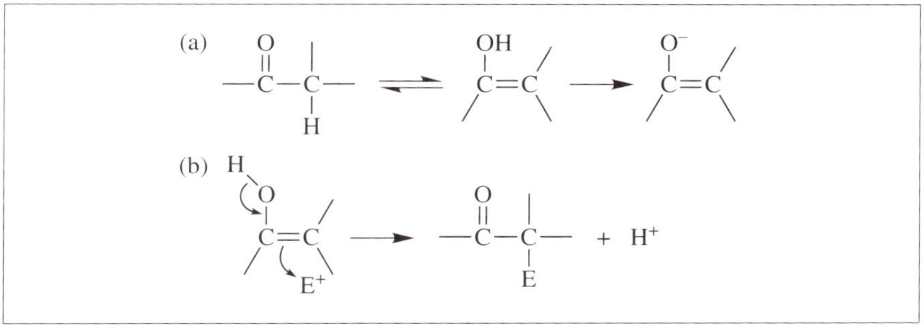

Scheme 3.35

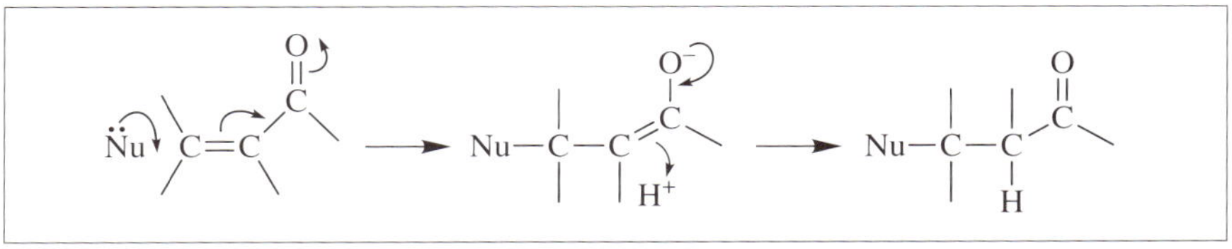

Scheme 3.36

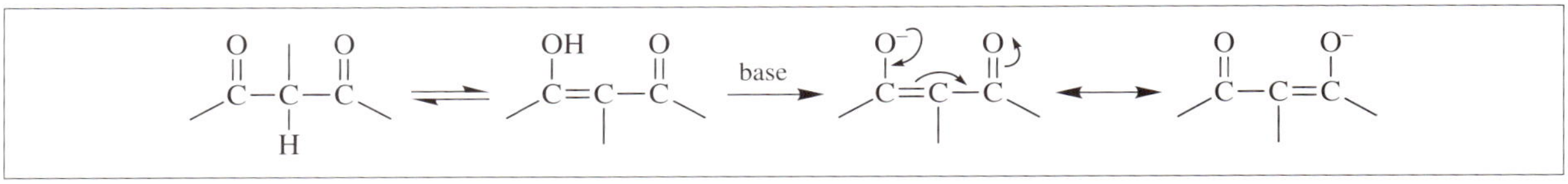

Scheme 3.37

The enolate of a β-dicarbonyl compound is acidic. The resultant anion may be neutralized by reaction with an electrophile either on the central carbon (Scheme 3.38a) or on an oxygen atom (Scheme 3.38b). This reactivity of β-dicarbonyl compounds makes them extremely useful in synthesis.

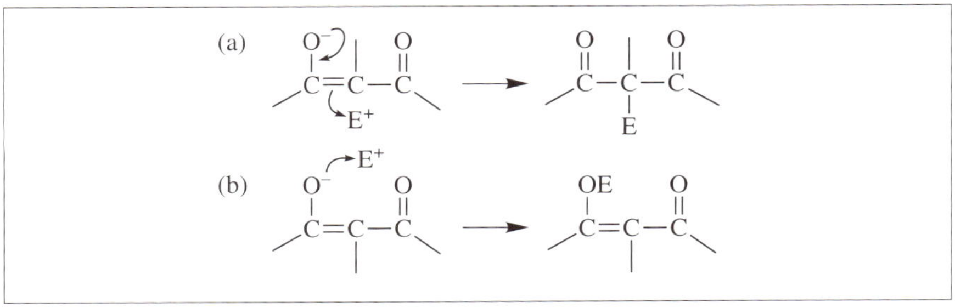

Scheme 3.38

The addition of a nucleophile to the carbonyl group involves the conversion of a planar sp^2 centre to a tetrahedral sp^3 centre with an increase in the steric bulk of the intermediate. The preferred direction of approach of the nucleophile to the carbonyl carbon is along an axis through the carbon and oxygen atoms and at an angle of 108° to the plane of the carbonyl group (see **3.7**).

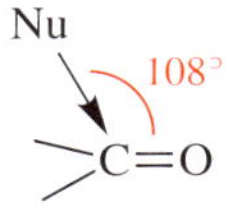

3.7

The face of the carbonyl group which is attacked by a nucleophile may be influenced by the size of the adjacent substituents. Several models have been proposed to account for the stereochemical outcome of these reactions. The validity of these models depends upon whether the reaction is a pure nucleophilic attack or is a reaction which is proceeding through catalysis by a Lewis acid and in which coordination to the oxygen precedes attack by the nucleophile. If the groups attached to the adjacent carbon atom are designated S (small), M (medium) and L (large), the optimum arrangement of these places the C–L bond perpendicular to the R–C=O fragment and the S group opposite the R alkyl residue (see **3.8**). A nucleophile is likely to approach the carbonyl group at an obtuse angle between the small and medium groups and in such a way as to minimize the interactions in the tetrahedral intermediate. Overall, this **Cram–Felkin–Ahn model** accounts for the majority of the results. In cyclic systems, more distant factors, such as the interactions with a methyl group, may affect the incoming nucleophile.

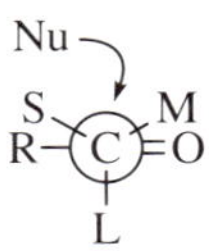

3.8

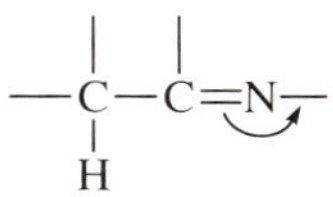

3.9

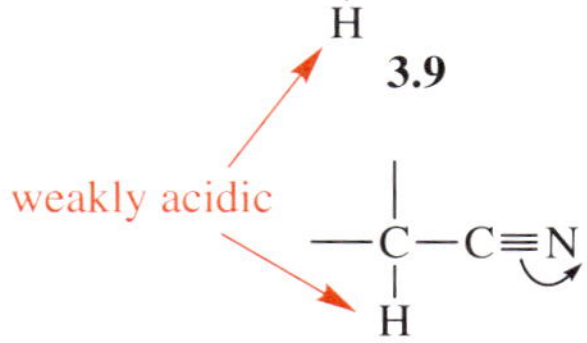

3.10

Carbonyl-like reactivity is found with a number of other functional groups. Replacement of the oxygen of the carbonyl (C=O) group with nitrogen leads to **imines** (C=N–R) and to **nitriles** (C≡N). These possess both an electron-deficient carbon and adjacent substituents which are activated by the imine (see **3.9**) or nitrile (see **3.10**).

Replacement of the carbon of the carbonyl group by nitrogen, phosphorus or sulfur reveals an analogy with nitroso and nitro compounds, phosphono derivatives, sulfoxides and sulfones (Box 3.3). Therefore it is not surprising to find parallels in the chemistry of these substances with those of the carbonyl group.

Box 3.3 Compounds Related to Carbonyls

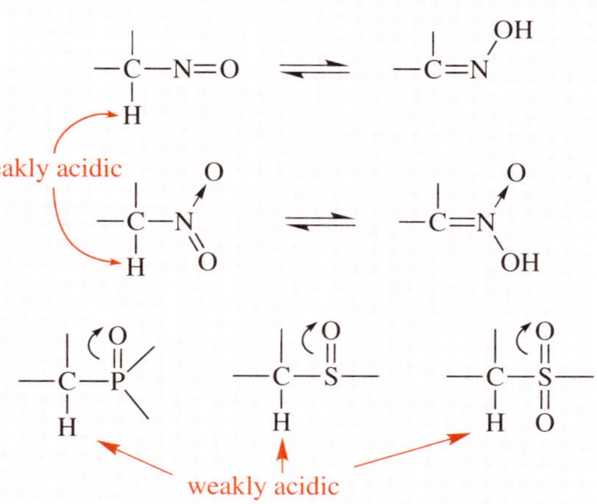

The reactions of the carbonyl group will be discussed under the headings of the general reactions of aldehydes and ketones and then the special reactions of aldehydes. Acyl chlorides, carboxylic acids, esters and amides will be discussed in the next section. The general reactions of the carbonyl group are shown in Box 3.4.

Box 3.4 Reactions of the Carbonyl Group

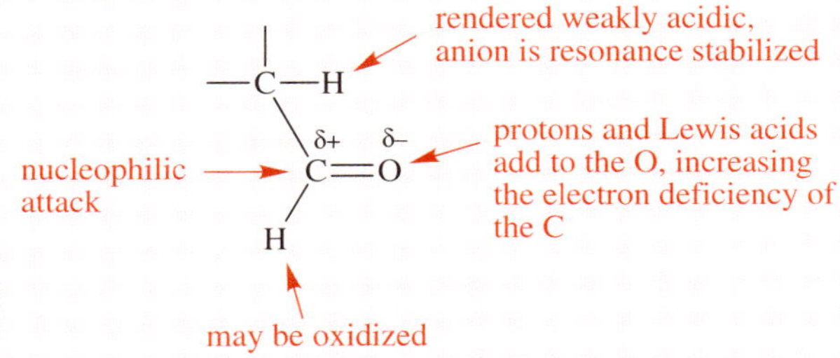

3.3.1 Reduction of Aldehydes and Ketones

Aldehydes and ketones may be reduced to the corresponding primary and secondary alcohols by reagents such as lithium aluminium hydride, sodium borohydride, sodium and ethanol or hydrogen over a platinum catalyst. A ketone is reduced to a methylene group under more vigorous conditions with zinc amalgam and concentrated hydrochloric acid (the **Clemmensen reduction**) or treatment of the hydrazone with alkali (the **Wolff–Kishner reduction**) (Scheme 3.39).

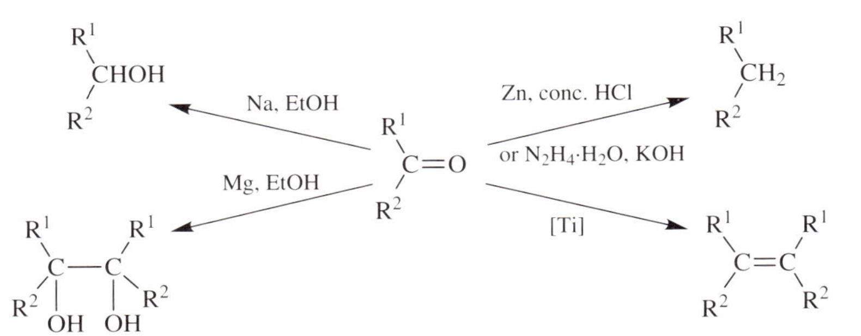

Scheme 3.39

Reagents such as magnesium, or low-valency states of titanium dissolving in acid, function by donating an electron to the carbonyl group to form a radical anion. The reductive process may be completed by the dimerization of these radicals to form 1,2-diols (pinacols) in the case of magnesium or alkenes in the case of titanium.

The **Meerwein–Ponndorf reduction** of ketones involves the transfer of hydrogen from an alcohol such as propan-2-ol (isopropanol) mediated by aluminium tri-isopropoxide. The isopropanol is oxidized to propanone (acetone). The reaction involves the intervention of a cyclic transition state (see Scheme 3.40) directed by the favourable formation of an O–Al bond. The reaction is driven to completion by the removal of the more volatile acetone by distillation.

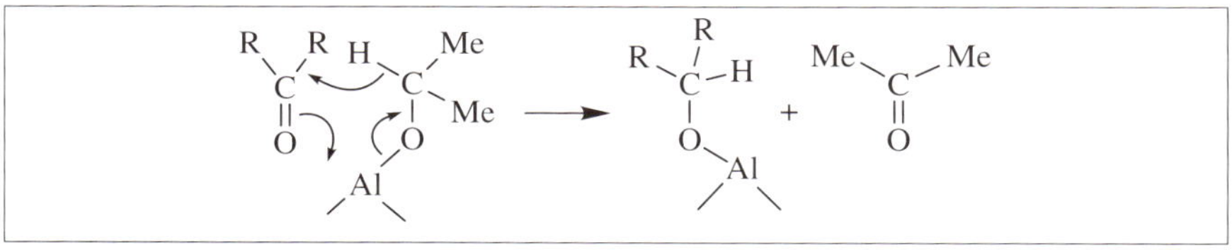

Scheme 3.40

The **Cannizzaro reaction** of non-enolizable aldehydes is another example of a hydride transfer reaction. It is carried out under alkaline conditions and involves not only the addition of a hydroxide ion to one aldehyde but the stabilization of the resultant acid as the anion (Scheme 3.41). Methanal (formaldehyde), which gives methanoic acid (formic acid), a relatively strong carboxylic acid, makes a good hydrogen donor in a cross-Cannizzaro reaction.

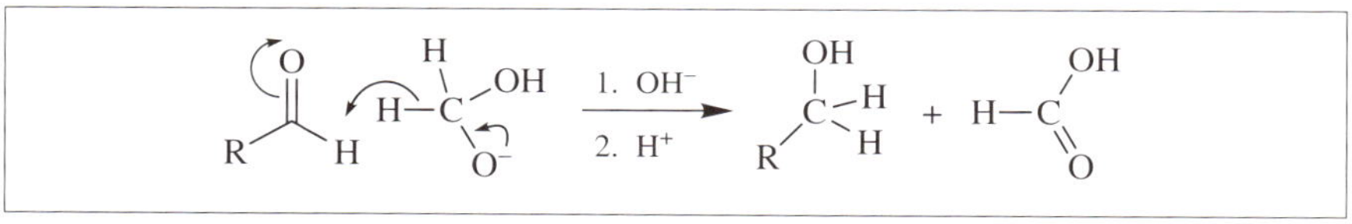

Scheme 3.41

3.3.2 Addition Reactions

Addition of Oxygen Nucleophiles

The addition of oxygen nucleophiles to the carbonyl group can take place quite easily, particularly under acid-catalysed conditions. A number of simple aldehydes such as formaldehyde and trichloroethanal (chloral) are extensively hydrated in aqueous solution. The addition of methanol to the carbonyl group under acid catalysis reveals a number of aspects of this chemistry (Scheme 3.42). The oxygen atom of the carbonyl group is protonated to produce an electron-deficient carbon atom. The initial addition takes place to form a **hemi-acetal**. In the presence of acid, water is lost from this with the formation of a carbocation that is stabilized by the lone pairs on the oxygen. Further addition of methanol then takes place with the formation of a **dimethyl acetal**. The formation of this tetrahedral centre leads to steric congestion in the molecule. Hence, although dimethyl acetals are formed quite easily from aldehydes, they are less readily formed by ketones, particularly sterically hindered ketones.

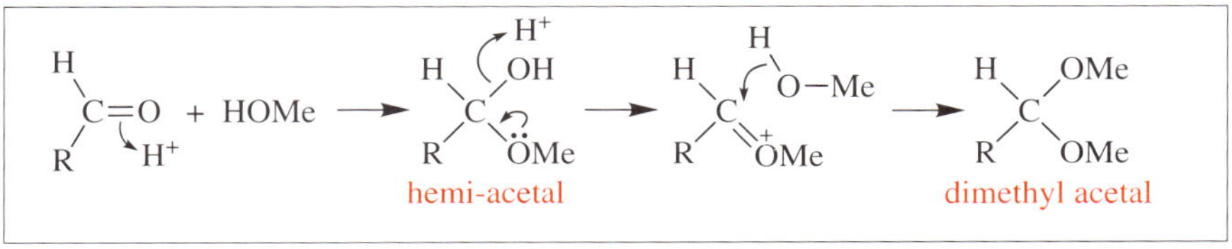

Scheme 3.42

Ethane-1,2-diol (ethylene glycol) reacts with aldehydes and ketones under these conditions to form ethylene acetals (**3.11**). These compounds no longer possess the electron-deficient carbon of the carbonyl group, and therefore these acetals function as **protecting groups** for the carbonyl group, allowing reactions to take place at other centres in the molecule without interference from the carbonyl group.

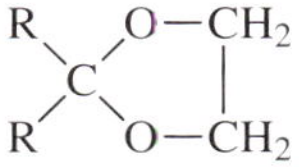

3.11

The oxygen nucleophile may be that of a peroxy acid such as perbenzoic acid. In this case a rearrangement may occur under acidic conditions. This takes place with the expulsion of benzoate and the insertion of an oxygen adjacent to the carbonyl group. The reaction, known as the **Baeyer–Villiger rearrangement**, leads to the conversion of a ketone to an ester (Scheme 3.43) or a cyclic ketone to a **lactone** (cyclic ester).

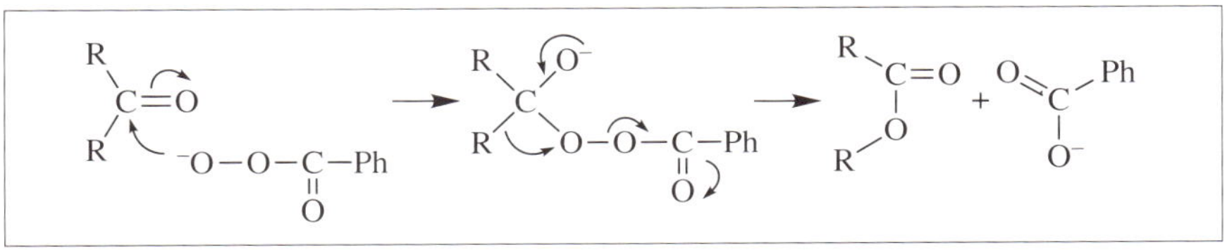

Scheme 3.43

Addition of Sulfur Nucleophiles

The addition of sulfur nucleophiles to aldehydes and ketones may be exemplified by the formation of hydrogensulfite (**bisulfite**) adducts (**3.12**). These are sulfonates that are water soluble. However, their steric bulk means that whereas they are formed from aldehydes and methyl ketones, more highly substituted ketones are reluctant to form these derivatives.

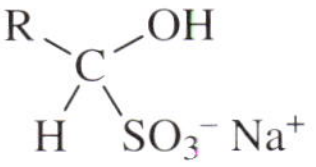

3.12

The formation of **thioacetals** from aldehydes and ketones involves the reaction with a thiol such as ethane-1,2-dithiol in the presence of a Lewis acid catalyst such as boron trifluoride etherate. These derivatives have been described earlier.

Addition of Nitrogen Nucleophiles

Amines behave as powerful nucleophiles and readily add to the carbonyl group. The initial addition product of the amine and the carbonyl

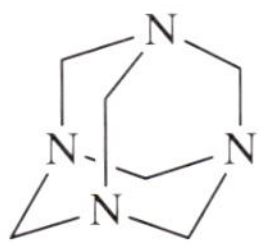

3.13

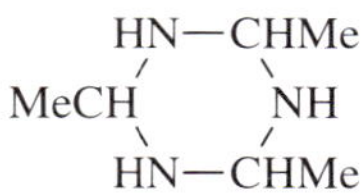

3.14

compound may undergo dehydration and the resultant imine may then be the substrate for a further addition.

Ammonia reacts with formaldehyde to give hexamethylenetetramine (**3.13**) and with acetaldehyde to give a trimer (**3.14**). Other more sterically hindered imines remain as the monoadduct and do not add a second molecule of amine. These imines are readily reduced by, for example, sodium cyanoborohydride to form amines or they may be hydrolysed to regenerate the carbonyl compound.

Secondary amines such as dimethylamine give rise to **iminium salts**. These salts, such as that formed from formaldehyde and dimethylamine, have a very reactive electron-deficient carbon. They are useful in C–C bond formation in the **Mannich reaction**.

The addition reactions of amines to polycarbonyl compounds are important ways of making heterocyclic compounds. An example is the condensation of ammonia with hexane-2,5-dione to give 2,5-dimethylpyrrole.

The reaction with hydroxylamine gives rise to **oximes** (Scheme 3.44a). The oximes of simple aldehydes and ketones are crystalline derivatives, the melting points of which play a useful role in characterizing these compounds. Oximes are planar, and show geometrical isomerism about the C=N bond. A useful feature of their chemistry is the **Beckmann rearrangement**. If the oximino hydroxyl group is converted into a leaving group by protonation or treatment with phosphorus pentachloride, a 1,2-shift of an alkyl or aryl group from carbon to nitrogen may occur with the ultimate conversion of a ketone to an amide. The stereochemistry of this reaction reflects the geometry of the oxime, with a *trans* relationship between the leaving hydroxyl group and the migrating C–C bond (Scheme 3.44b).

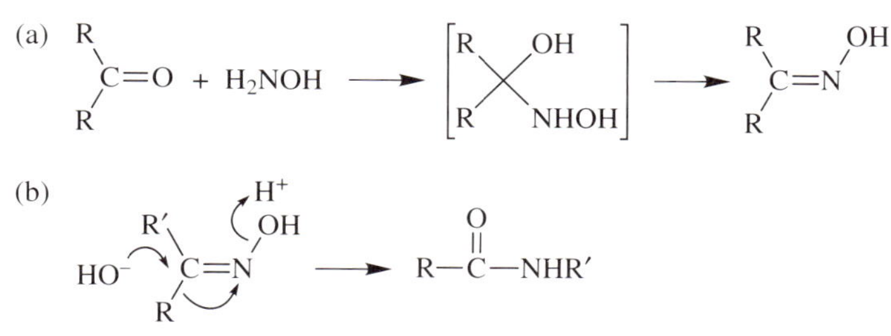

Scheme 3.44

The addition reactions of **hydrazines** provide a number of useful solid derivatives of well-defined melting point, such as the phenylhydrazones, the 2,4-dinitrophenylhydrazones (Scheme 3.45a) and the semicarbazones (Scheme 3.45b). These are used to characterize aldehydes and ketones. In each case the reaction takes place at the most nucleophilic nitrogen.

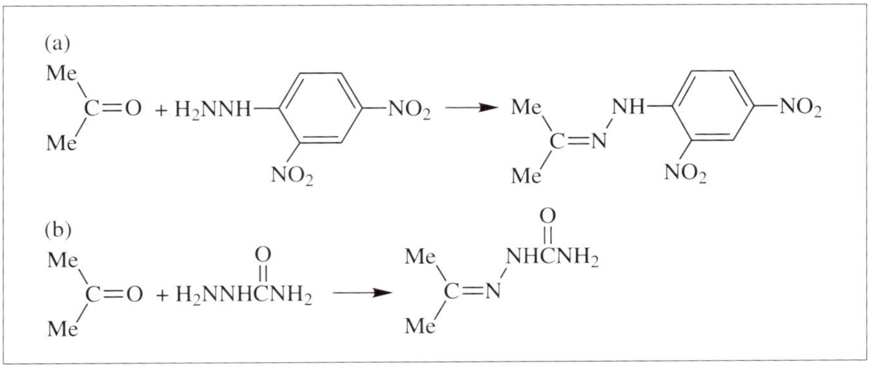

Scheme 3.45

A useful reduction reaction takes place with hydrazine in the presence of a strong base. In the Wolff–Kishner reduction a carbonyl group is reduced to a methylene. If the hydrazone is formed from an epoxy ketone, the decomposition takes place easily with the formation of an allylic alcohol.

Addition of Carbon Nucleophiles

The reaction of carbon nucleophiles with the electron-deficient carbon of a carbonyl group represents one of the major ways of making C–C bonds. The addition of hydrogen cyanide to acetone to form the cyanohydrin (*e.g.* **3.15**) was one of the first reactions to be studied mechanistically. The reversible reaction leads to the cyanohydrin in which the cyano group may be further modified by hydrolysis to a carboxylic acid or by reduction to an amine.

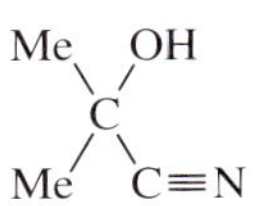

3.15

An amino nitrile may be formed instead of a cyanohydrin if the carbonyl compound is treated with aqueous ammonium chloride and sodium cyanide. Hydrolysis of the amino nitrile gives an α-amino acid as in the Strecker synthesis of DL-alanine (Scheme 3.46).

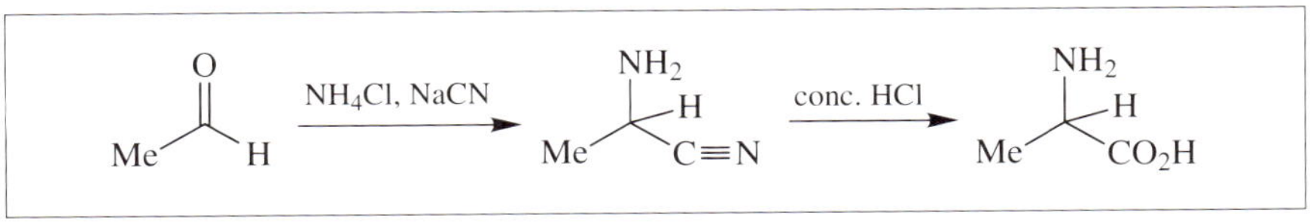

Scheme 3.46

The nucleophilic addition of an acetylide to form an ethynyl alcohol and the addition of organometallic reagents such as Grignard reagents to carbonyl compounds are important methods for C–C bond formation.

Useful carbon nucleophiles are derived from the phosphorus, sulfur and silicon ylides. The elimination of a hydrogen halide from a phosphonium salt by a strong base, such as sodium hydride, leads to the formation of a carbanion (see Scheme 3.47). The carbanion of this ylide is stabilized by the adjacent positively charged phosphorus in these **Wittig reagents**. The addition of these carbanions to a carbonyl group generates a dipolar intermediate, a betaine. Decomposition of this adduct, with the formation of triphenylphosphine oxide, brings about the regiospecific formation of an alkene from a carbonyl group.

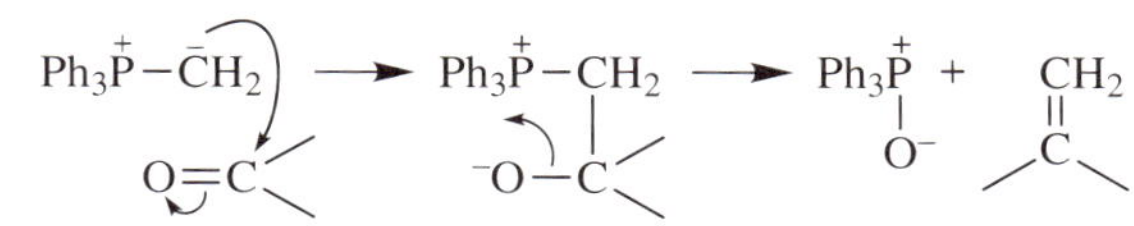

Scheme 3.47

The phosphorus ylide may be obtained from a wide range of different halides, and thus ketones may be converted into a range of unsaturated compounds. A useful reaction with the methoxymethylene Wittig reagent leads, via the acid-labile enol ether, to an aldehyde (Scheme 3.48).

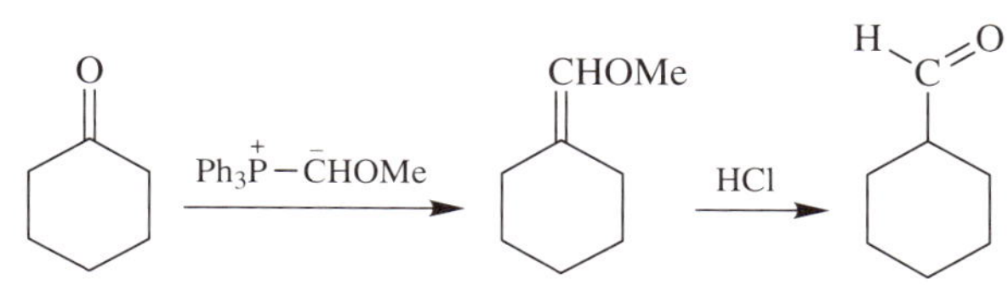

Scheme 3.48

The use of sulfur in place of the phosphorus brings about a different mode of decomposition of the intermediate betaine. Two sulfur ylides, dimethylsulfonium methylide (Scheme 3.49a) and dimethylsulfoxonium methylide (Scheme 3.49b), have been used. Both ylides react with ketones to give epoxides, but the stereochemistry may differ.

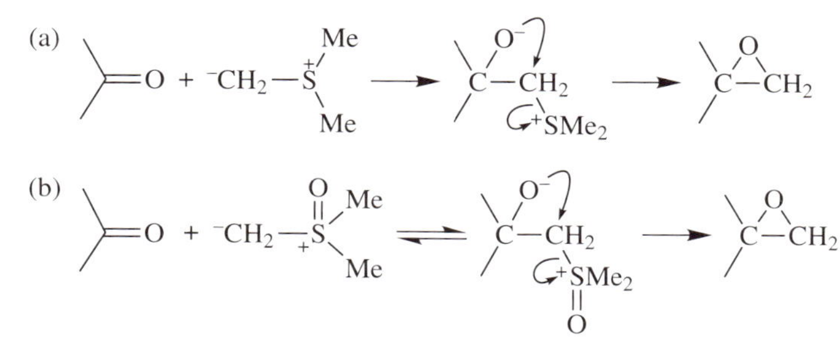

Scheme 3.49

The use of silicon is illustrated by Scheme 3.50 involving a Grignard reagent containing a trimethylsilyl group in which the silicon stabilizes the carbanion. Elimination of the silyl group from the β-hydroxysilane leads to the formation of the alkene.

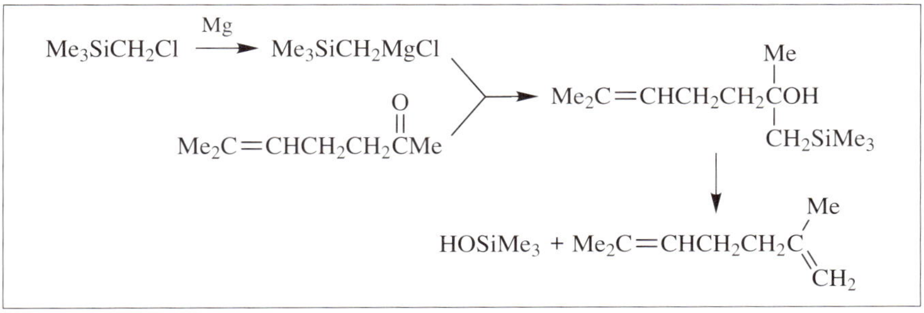

Scheme 3.50

The carbonyl group acts as the recipient carbon for a large family of carbanions derived from carbonyl activation. These are dealt with in more detail in the section on enolate anions.

3.3.3 Special Properties of Aldehydes

Aldehydes show a number of special properties which distinguish them from ketones. Firstly, they are easily oxidized to the corresponding carboxylic acid. If alkaline silver oxide is used for this purpose, a silver mirror is formed, and this can be used as a test for an aldehyde.

The oxidation of aldehydes by reagents such as chromium(VI) oxide involves the hydration of the aldehyde and hence sterically hindered aldehydes are sometimes relatively slow to be oxidized. Many aldehydes are autoxidized *via* the hydroperoxide to give the acid or other decomposition products.

Non-enolizable aromatic aldehydes take part in the Cannizzaro reaction.

The cyanohydrins of aromatic aldehydes participate in the benzoin condensation. The carbon-bound hydrogen (formerly the aldehydic hydrogen) of the cyanohydrin is made weakly acidic by the adjacent nitrile. The related carbanion may condense with a second molecule of the aldehyde. The resultant cyanohydrin is that of a ketone which is less stable and collapses to give benzoin (Scheme 3.51).

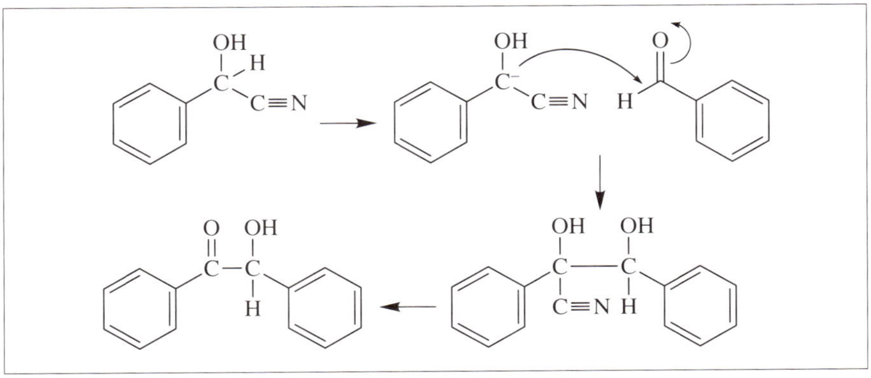

Scheme 3.51

3.3.4 Quinones

Quinones are cyclic unsaturated diketones in which the carbon atoms are derived from the oxidation of an aromatic system. Although the underlying reactivity of quinones is that of unsaturated ketones (cyclohexadienediones), it is tempered by this relationship to aromatic compounds and in particular by their reduction to dihydroxyphenols (quinols).

The relationship of quinones to aromatic compounds is revealed by their preparation by the oxidation of aromatic hydrocarbons, phenols or amines (Scheme 3.52).

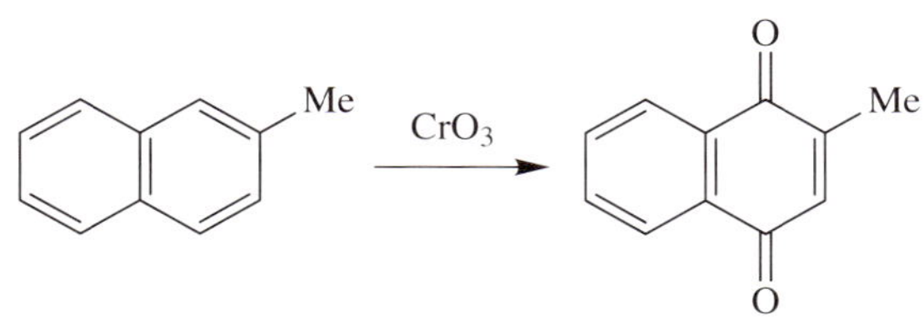

Scheme 3.52

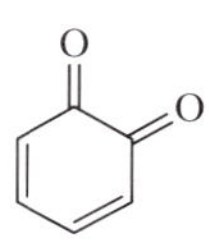

3.16 *o*-quinone

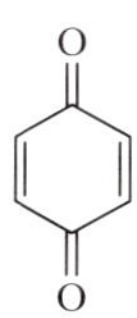

3.17 *p*-quinone

Two major types of quinone exist, 1,2- and 1,4-quinones, exemplified by *ortho*- (**3.16**) and *para*-benzoquinone (**3.17**). Quinones are electron-deficient conjugated systems which can accept electrons in two single-electron steps to give, firstly, a semiquinone radical anion and then a hydroquinone (Scheme 3.53). The electrode potential required to bring about the change varies with the structure. Electron-withdrawing substituents increase the electron deficiency and make the quinone a more powerful electron acceptor and consequently oxidizing agent. Tetrachloro-*p*-benzoquinone (chloranil) and 2,3-dichloro-4,5-dicyano-*p*-benzoquinone are useful dehydrogenating agents for converting ketones to unsaturated ketones.

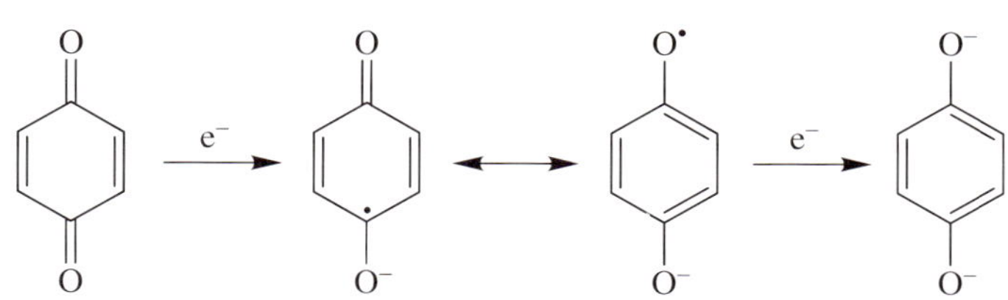

Scheme 3.53

Quinones are reduced to quinols by sulfur dioxide or by zinc dust in acetic acid. The ability of a quinone to act as an electron-acceptor and their reduction products (quinols) to act as electron-donors is reflected in the formation of quinhydrones which are highly coloured 1:1 charge-transfer molecular complexes.

Quinones behave as typical unsaturated ketones and undergo 1,4-addition reactions with acid catalysis. Thus treatment of *p*-benzoquinone with hydrogen chloride gives a chlorohydroquinone (**3.18**). The addition of aniline and methanol takes place to give initially the corresponding quinols. However, the electron-donating character of these substituents makes the quinols more easily oxidized by excess *p*-benzoquinone. The anilino- (**3.19**) and methoxyquinones (**3.20**) are therefore the final products.

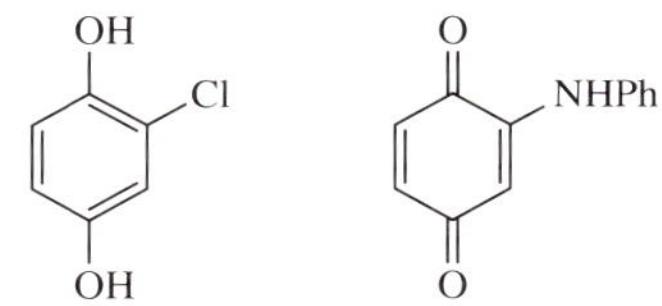

3.18 **3.19**

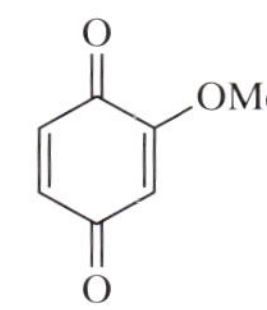

3.20

Quinones, typical of unsaturated ketones, undergo base-catalysed epoxidation. The epoxy ketone may then rearrange with the formation of a hydroxy quinone. Quinones may also act as dienophiles in the Diels–Alder reaction. This can be a useful way of constructing polycyclic ring systems (Scheme 3.54).

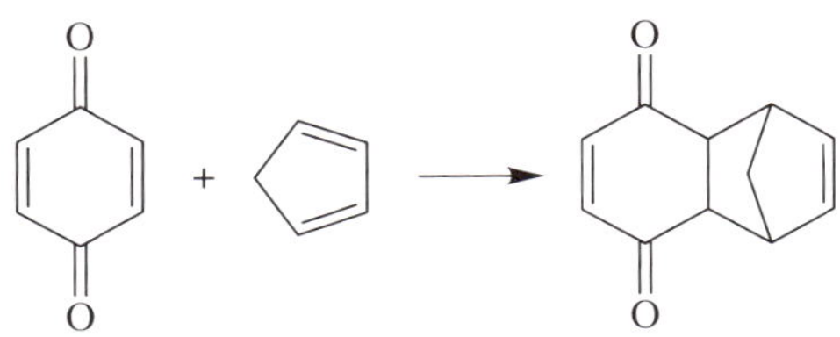

Scheme 3.54

Exercise

Copy this revision chart showing the relationship between aldehydes and ketones and other functional groups, and fill in the relevant reagents and conditions for the inter-relationships beside the arrows.

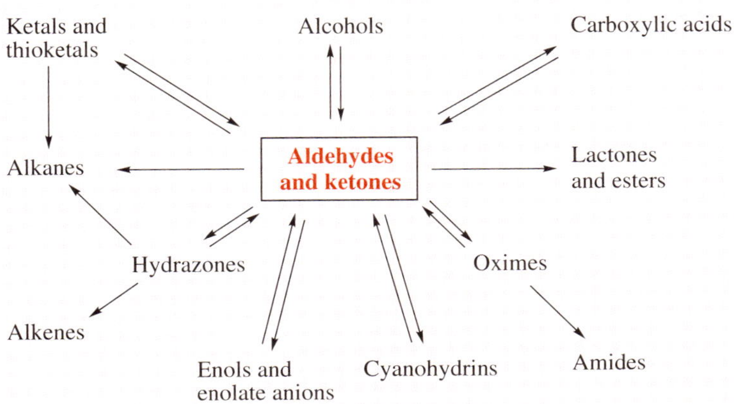

3.4 Carboxylic Acids and their Relatives

The carbonyl group has a major effect on the reactivity of a number of substituents such as the halogen of an acyl halide, the hydroxyl group of a carboxylic acid and the amino group of an amide. Although it is possible to regard each of these as separate functional groups, in each case the carbonyl group modifies the properties of the substituent, while the substituent also modifies the properties of the carbonyl group. There are some similarities in these interactions which are usefully considered together.

3.4.1 Carboxylic Acids

The electron-withdrawing properties of the carbonyl group renders acidic the hydrogen of the attached hydroxyl group. The **carboxylate anion** is then stabilized by resonance (Scheme 3.55). The strength of the individual carboxylic acids is a reflection of the other substituents attached to the carbonyl group. A powerful electron-withdrawing group, such as the trifluoromethyl group, increases the strength of the acid, while an electron-donating substituent weakens the acid.

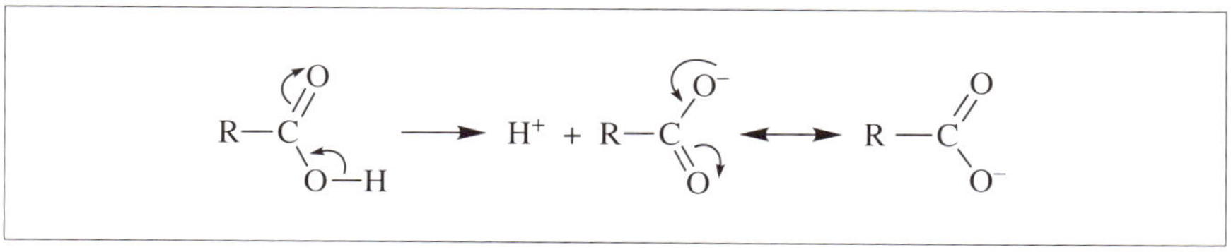

Scheme 3.55

Reduction of Carboxylic Acids

The use of **hydride reagents** for the reduction of carboxylic acids requires vigorous conditions because the carboxylate anion, once formed, is resistant to further nucleophilic attack. An interesting exception to this is reduction with borane. A triacyloxyborane may be formed. Resonance involving the vacant orbitals of the boron and the lone pairs on the oxygen has the effect of making the attack of a nucleophilic hydride on the carbonyl carbon of the acid easier (see **3.21**).

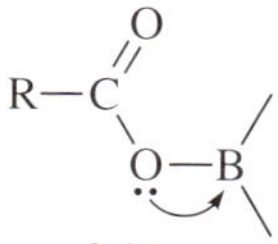

3.21

3.4.2 Formation of Acyl Derivatives

The key to the formation of acyl derivatives involves removing the influence of the oxygen lone pairs on the electron-deficient carbon atom of the carbonyl group. This may be achieved either by protonation or by the formation of a mixed anhydride. Thus **esterification** occurs with an alcohol and an acid catalyst (Scheme 3.56). Protonation of the carbonyl group by the acid catalyst facilitates the addition of the alcohol to the carbonyl carbon atom, and leads to the formation of a tetrahedral intermediate. The loss of water from this intermediate gives the ester.

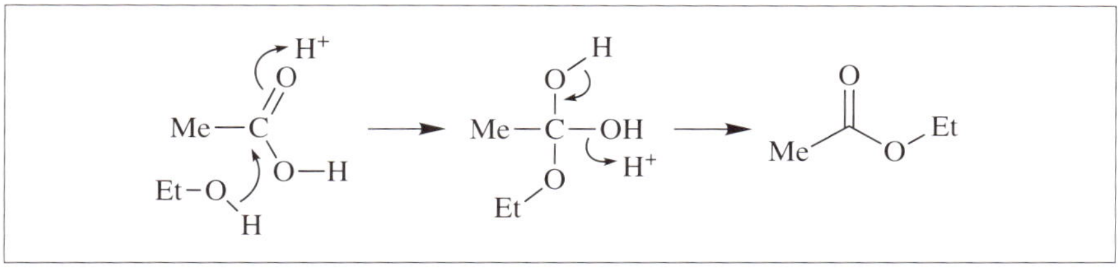

Scheme 3.56

Reaction of the acid with **diazomethane** (CH_2N_2) gives methyl esters. Diazomethane is a 1,3-dipolar molecule, the structure of which can be written in a number of resonance forms (Scheme 3.57a). The reaction with a carboxylic acid is accompanied by the loss of nitrogen gas (Scheme 3.57b).

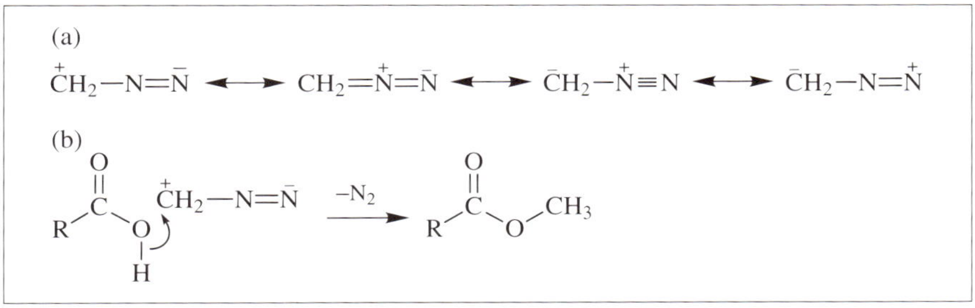

Scheme 3.57

A carboxylic acid may be converted to an **acid anhydride** by heating the acyl chloride with a salt of the same acid (Scheme 3.58). Acid anhydrides may also be prepared by heating the acid with acetic anhydride.

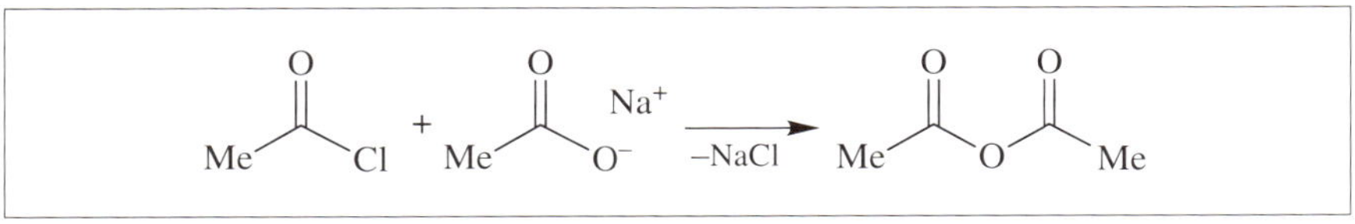

Scheme 3.58

Acid anhydrides react more easily than the corresponding carboxylic acid with nucleophiles because the electron-withdrawing effect of the second carbonyl group reduces the effect of the lone pairs on the oxygen on the first carbonyl group (see **3.22**).

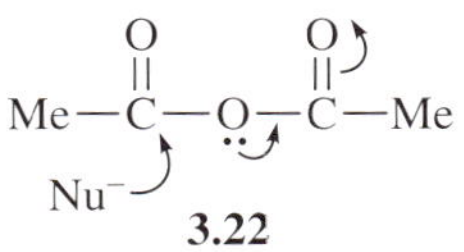

3.22

Once the tetrahedral intermediate has been formed by addition of the nucleophile to a carbonyl, the second carboxyl group acts as a good leaving group, favouring the collapse of the intermediate and the formation of the product. Acid anhydrides are therefore widely used in making the esters of alcohols.

The formation of acylium ions by the fission of the hydroxyl group in strong acid provides a useful source of electrophilic carbon and this leads to an important series of methods for making C–C bonds (Scheme 3.59), similar to the Friedel–Crafts acylations based on acyl chlorides. The acylium ion obtains some stabilization from the lone pairs on the oxygen atom of the carbonyl group (see **3.23**).

3.23

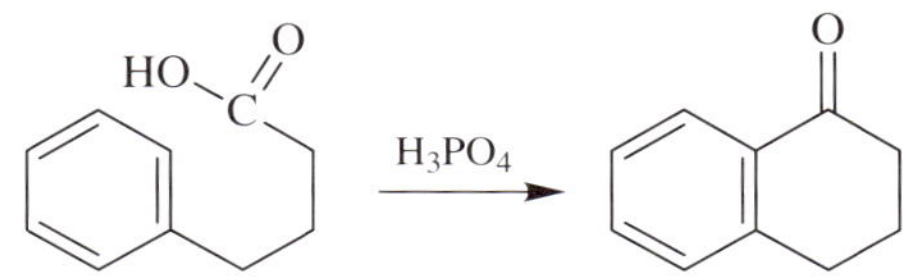

Scheme 3.59

The carboxylate anion is a useful oxygen nucleophile for the displacement of a variety of leaving groups. A halophilic silver salt may be used for the displacement of a halogen such as an alkyl halide.

The carboxylate anion may neutralize a carbocation formed by the attack of an electrophilic metal on an alkene, as in the oxymercuration of alkenes.

The decarboxylation of acids may take place by both radical and ionic processes. Radical processes involving the electrolytic discharge of a carboxylate anion (the Kolbe reaction) may give rise to dimeric products.

Silver salts of carboxylic acids undergo a halodecarboxylation in the presence of bromine (the Hunsdiecker reaction, Scheme 3.60).

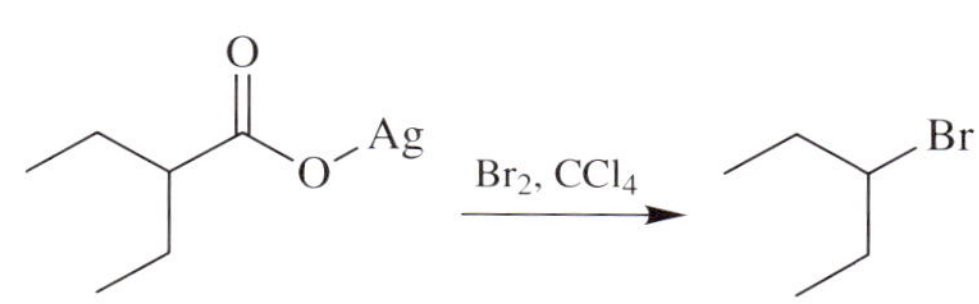

Scheme 3.60

Decarboxylation of the aliphatic esters of *N*-hydroxypyridine-2-thione by tributyltin hydride leads to reduction of the radical and the formation of the alkane lacking the carboxyl carbon, as in the Barton–McCombie reaction (Scheme 3.61).

A number of reactions of metal salts can be rationalized in terms of the formation of a carbanion adjacent to the carboxylate. Dibasic metals such as calcium bring two carboxylate units close to each other so that the carbanion formed adjacent to one carboxylate may attack the carbonyl of the other. Thus pyrolysis of calcium acetate affords propanone (acetone) (Scheme 3.62). A similar reaction is found in the pyrolytic cyclization of some dicarboxylic acid anhydrides. Heating C_6 and C_7 dicarboxylic acids gives cyclopentanones and cyclohexanones

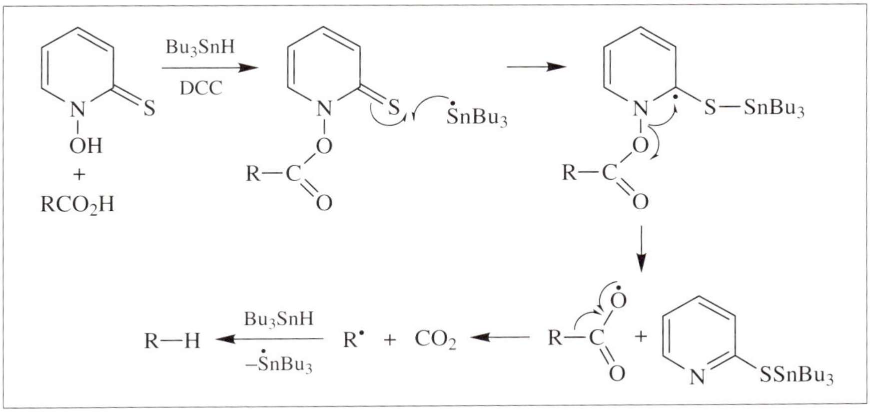

Scheme 3.61

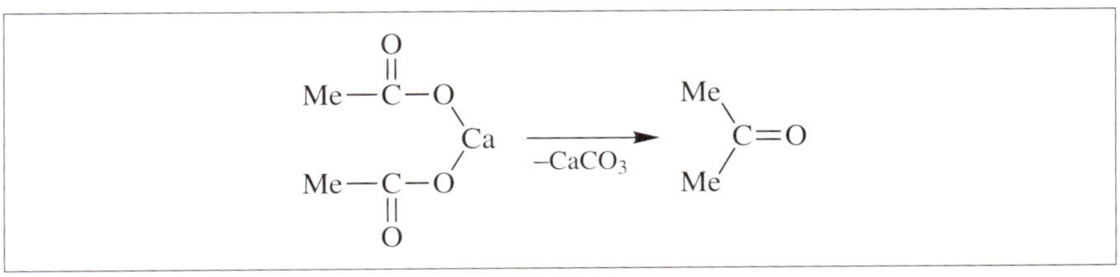

Scheme 3.62

respectively, but smaller rings such as cyclobutanones cannot be formed this way.

3.4.3 Acyl Halides

Acyl halides may be prepared from carboxylic acids by treatment with phosphorus pentachloride or, better, thionyl chloride (Scheme 3.63); the latter gives gaseous by-products. Oxalyl chloride may be used in the same way and again it gives volatile by-products.

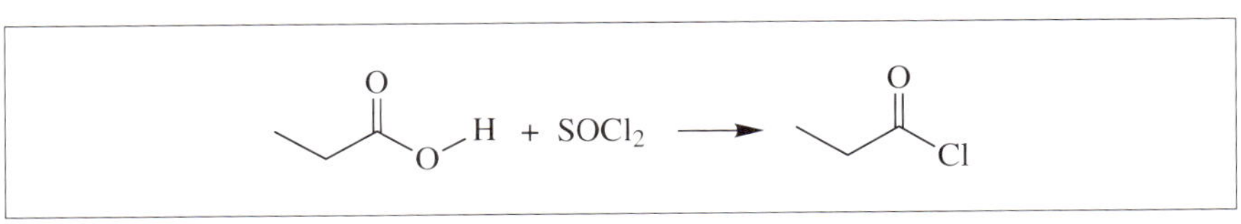

Scheme 3.63

The electron-withdrawing effect of the halogen, coupled with that of the carbonyl oxygen, leads to a very electron-deficient carbon, and this is not effectively counteracted by the lone pairs on halogens such as chlorine. Consequently, the carbonyl carbon atom is very sensitive to nucleophilic addition to form a tetrahedral intermediate. The collapse of the tetrahedral intermediate with the expulsion of the halide ion, which is a good leaving group, enhances the reactivity of the acyl halides (Scheme 3.64a). The direct fission of the acyl halide C–X bond leads to the formation of an electrophilic **acylium ion** (Scheme 3.64b).

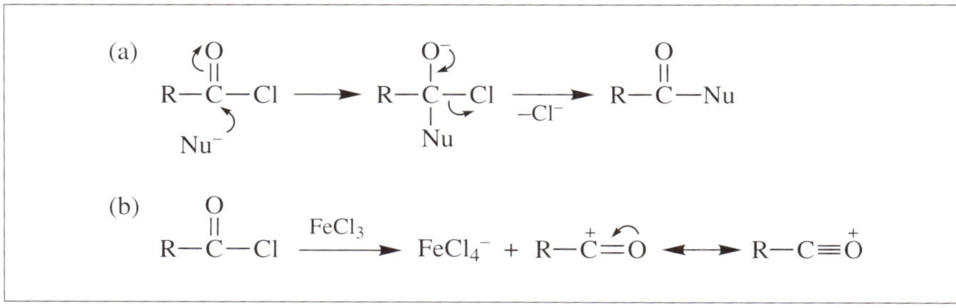

Scheme 3.64

Acyl halides are sensitive to the same groups of reducing agents that are used with aldehydes and ketones. However, it is possible, by using a catalyst poison such as barium sulfate, to stop the reduction at the stage at which only the halogen atom has been replaced by hydrogen, as in the **Rosenmund reduction** (Scheme 3.65).

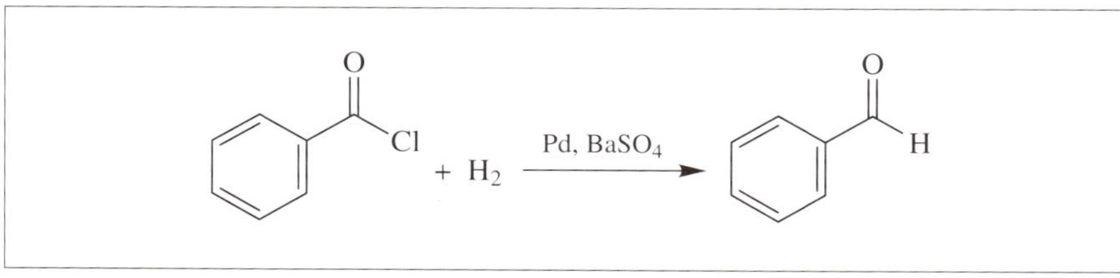

Scheme 3.65

Acyl halides such as acetyl chloride react with water to regenerate the starting acid and they react with alcohols to yield **esters**. The reaction is often facilitated by the presence of a tertiary amine catalyst. Although this may function purely as a base to neutralize the hydrochloric acid which is formed in the reaction, bases such as **dimethylaminopyridine** can also activate the carboxyl group via the formation of the intermediate shown in Scheme 3.66.

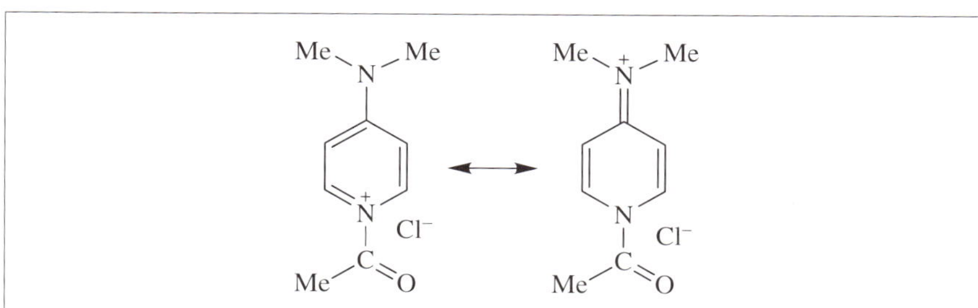

Scheme 3.66

The **Schotten–Baumann esterification** procedure is used with benzoyl chloride and other aroyl chlorides that are relatively stable to water. In this procedure the alcohol is shaken with the acyl chloride in the presence of aqueous sodium hydroxide. This method is particularly useful with phenols (Scheme 3.67).

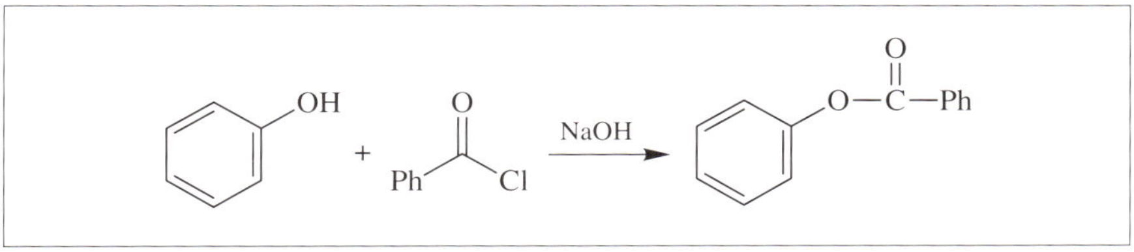

Scheme 3.67

Nitrogen nucleophiles such as ammonia and amines react rapidly with acyl chlorides to form amides.

Acyl chlorides react with **organometallic reagents**. Ketones may be formed when reagents such as organocadmium compounds are used (Scheme 3.68).

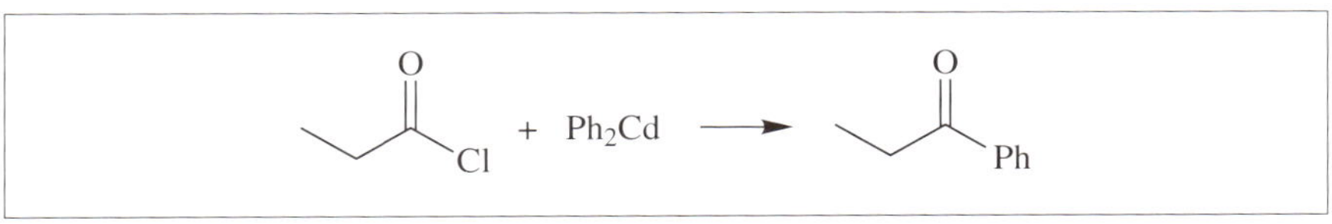

Scheme 3.68

A useful reaction of acyl chlorides is with diazomethane. The initial product, a diazo ketone, undergoes a rearrangement with the expulsion of nitrogen and the insertion of a methylene group. This **Arndt–Eistert reaction** is a useful chain-extension procedure (Scheme 3.69).

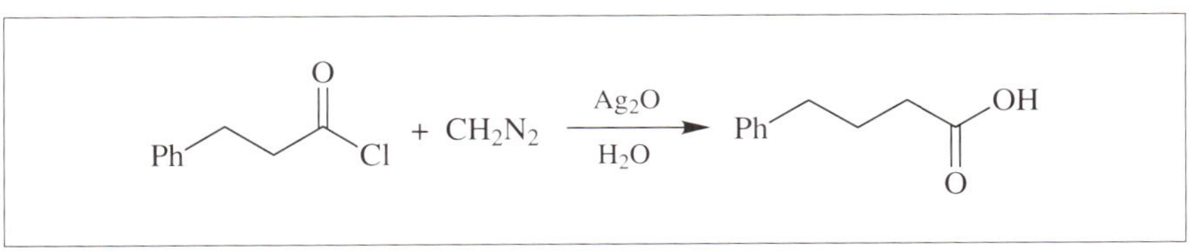

Scheme 3.69

3.4.4 Reactions of Esters

Esters are more readily reduced than carboxylic acids because there are not the problems associated with carboxylate formation. Thus reduction with lithium aluminium hydride gives a primary alcohol from the carboxylic acid component (Scheme 3.70).

The hydrolysis of esters may take place by two general pathways involving either **acyl–oxygen fission** (Scheme 3.71a) or **alkyl–oxygen**

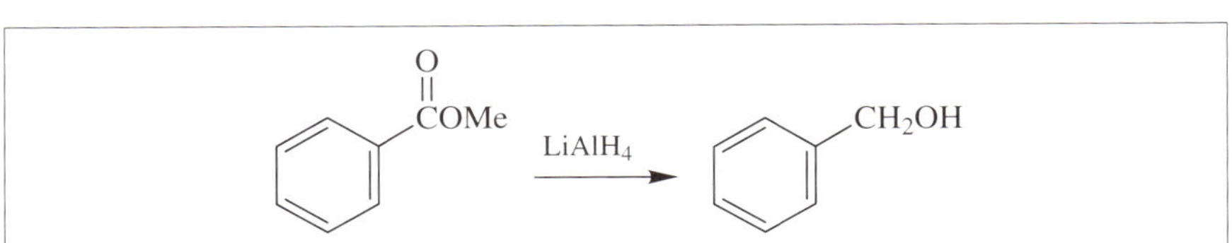

Scheme 3.70

fission (Scheme 3.71b). Under normal basic conditions a tetrahedral intermediate is formed by the addition of a nucleophile to the carbonyl group. This collapses with the expulsion of an alkoxide. Under acid-catalysed conditions the initial protonation of the carbonyl group renders the carbonyl carbon more sensitive to attack by a nucleophile. When the carbonyl group is particularly hindered, attack on the alkyl ester carbon may take place. This may be facilitated by a hard acid–soft base pair such as trimethylsilyl iodide (from trimethylsilyl chloride and sodium iodide). The hard silicon bonds to the carbonyl oxygen, increasing the electron deficiency of the carbonyl carbon, whilst the soft iodide attacks the ester carbon. Very dry lithium iodide can be used for the same purpose.

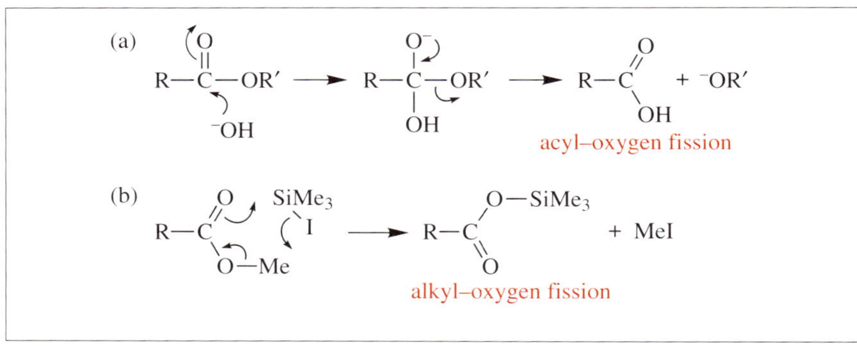

Scheme 3.71

Because the hydroxyl hydrogen is replaced by an alkyl group, the acidifying effect of the carbonyl group on an adjacent C–H bond can be expressed. A carbanion may be formed in the presence of a base such as sodium ethoxide or a bulkier group such as potassium *t*-butoxide that does not form a tetrahedral intermediate by addition to the carbonyl group. These carbanions may then react as nucleophiles by addition to other carbonyl groups in condensation reactions or to displace the halogen in alkyl halides in alkylation reactions.

3.4.5 Reactions of Amides

The chemistry of amides is characterized by diminished carbonyl reactivity and a diminished basicity and nucleophilicity of the nitrogen, arising from the interaction between the carbonyl group and the nitrogen lone pairs (Scheme 3.72a). An understanding of the chemistry of amides is important because of the role that the amide group plays in the peptide bond. On the other hand, the carbonyl group renders the amide hydrogen atoms weakly acidic and, in the presence of strong bases, anions may be formed (Scheme 3.72b).

The reduction of amides to amines may be achieved with lithium aluminium hydride and with borane (Scheme 3.73).

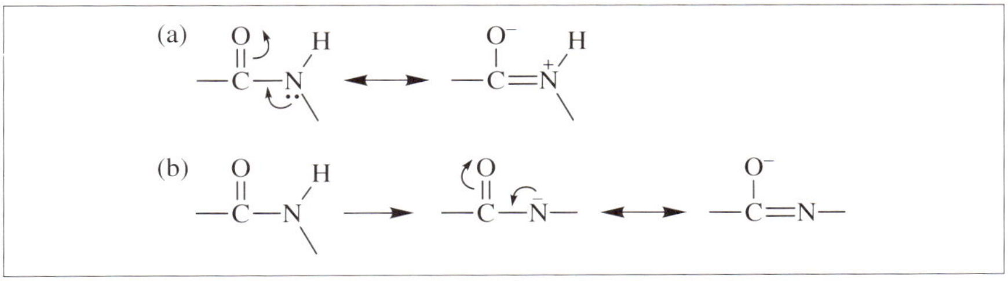

Scheme 3.72

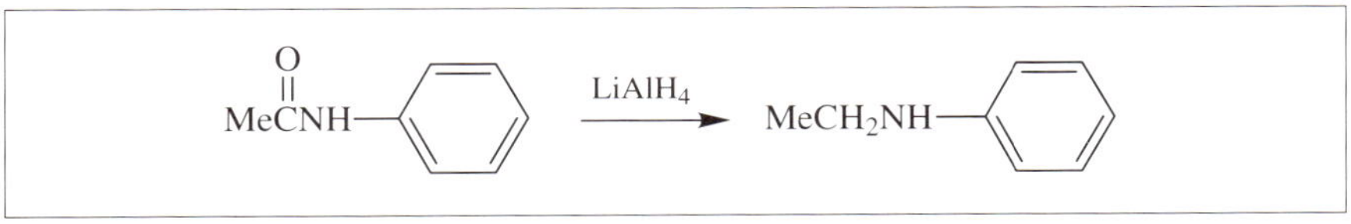

Scheme 3.73

The hydrolysis of amides is normally carried out under acidic catalysis. Protonation makes the carbonyl group more sensitive to nucleophilic attack (see **3.24**).

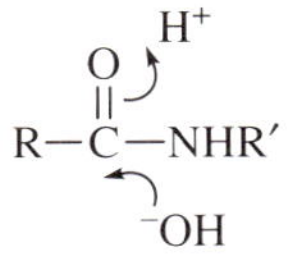

3.24

The weakly acidic N–H may be **alkylated** in the presence of a base. This can provide a useful method for making amines.

The amide N–H may also be halogenated, oxidized and nitrosated. *N*-Bromosuccinimide (NBS), like a number of other *N*-halo compounds, readily undergoes a radical fission to give a bromine radical. This provides a useful reagent for radical bromination at, for example, allylic or benzylic positions. In the presence of acid, NBS is also a mild source of the halonium ion, which is used for the addition of hypobromous acid (Scheme 3.74) to alkenes or for the bromination of reactive aromatic rings.

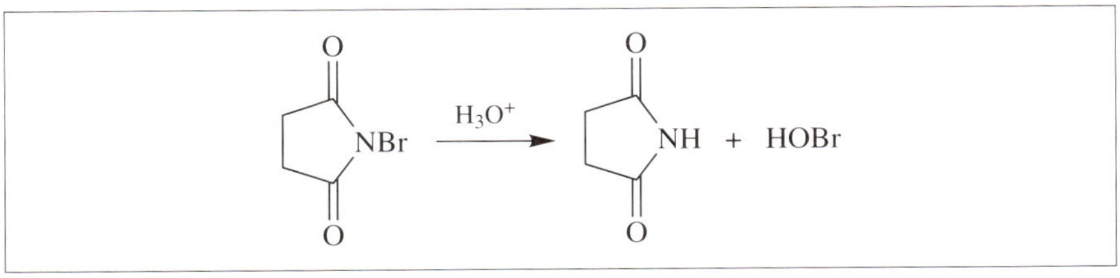

Scheme 3.74

N-Haloamides also undergo the **Hofmann rearrangement**. This reaction provides a useful method of making amines from amides with the loss of the carboxyl carbon atom.

Exercise

Copy this revision chart showing the relationship between carboxylic acids and other functional groups, and fill in the relevant reagents and conditions for the inter-relationships beside the arrows.

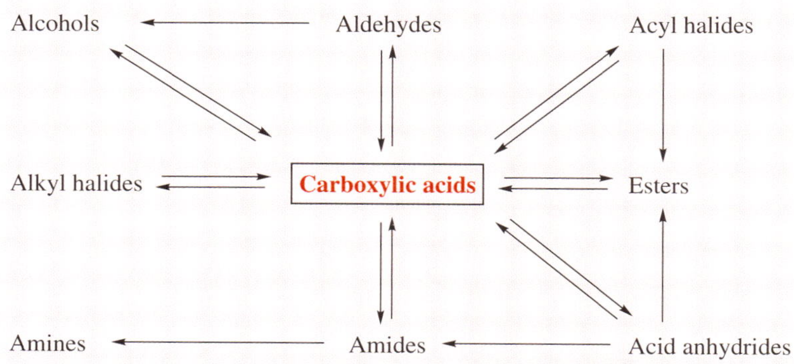

3.5 Enolate and Related Carbanion Chemistry

3.5.1 Enolate Anion Formation and Stabilization

The electron-withdrawing effect of a carbonyl group makes the hydrogen atom of an adjacent C–H weakly acidic. Once the anion is formed, it is stabilized by resonance. The negative charge is delocalized over both the carbonyl oxygen and this carbon. The enolate anion can react as a nucleophile at either the carbon or the oxygen (Scheme 3.75).

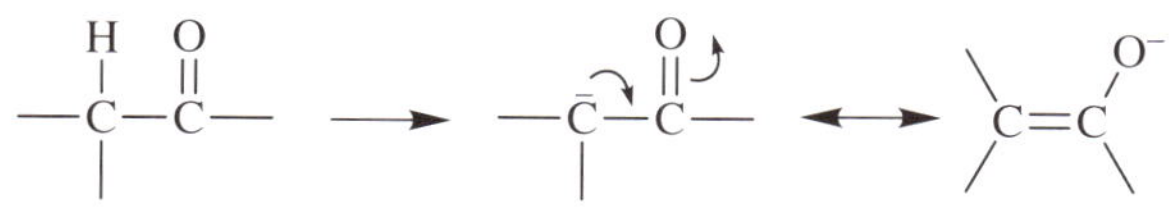

Scheme 3.75

The acidity of the C–H varies according to the type of carbonyl compound. Thus, because of the electron donation from the ester oxygen lone pair to the carbonyl group, the C–H adjacent to an ester is less acidic than that of a ketone. On the other hand, the C–H lying between two carbonyl groups is considerably more acidic than a C–H adjacent to just one carbonyl group (Scheme 3.76).

A number of other groups such as the nitrile, imine, nitro, sulfoxide, sulfonyl and phosphono, can act as **activating groups**. These groups may

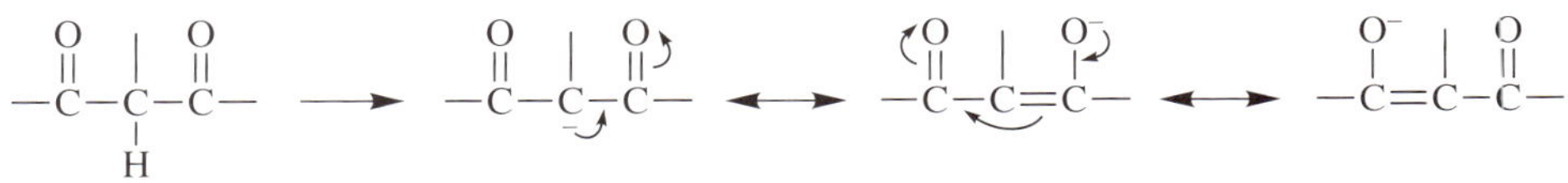

Scheme 3.76

be used either on their own or in combination with other carbonyl groups to generate a range of synthetically useful **carbanions**.

In each of the cases where there are two activating groups, the anions are stabilized by delocalization over both activating groups. Different bases are required to generate the anions from these acidic hydrogens.

An unsymmetrical ketone can form two different enolates. In some situations it is possible to distinguish between them by trapping the separate enolates as their silyl enol ethers. The anions may then be regenerated from the silyl enol ether in an aprotic solvent under non-equilibrating conditions using fluoride ion. The rapidly formed kinetic enol of 2-methylcyclohexanone may be trapped using lithium diisopropylamide as the base (Scheme 3.77a). On the other hand, the thermodynamically more stable enol is trapped with a milder base such as triethylamine (Scheme 3.77b).

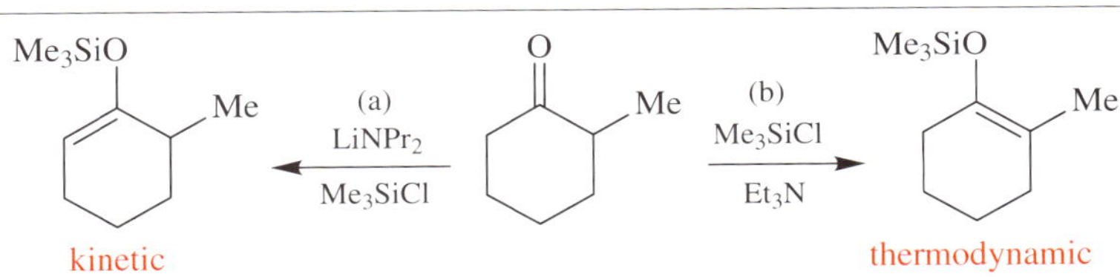

Scheme 3.77

Protonation of an enolate anion may take place on the oxygen or on the carbon. This may lead to the exchange of a proton by deuterium. The formation of an enolate and the subsequent regeneration of a ketone may lead to **epimerization** and, in the case of a chiral centre, to **racemization**

3.5.2 Reactions of Enolate Anions

Alkylation adjacent to a ketone is a useful synthetic reaction (Scheme 3.78). The reaction involves the nucleophilic substitution of the alkyl halide by the carbanion.

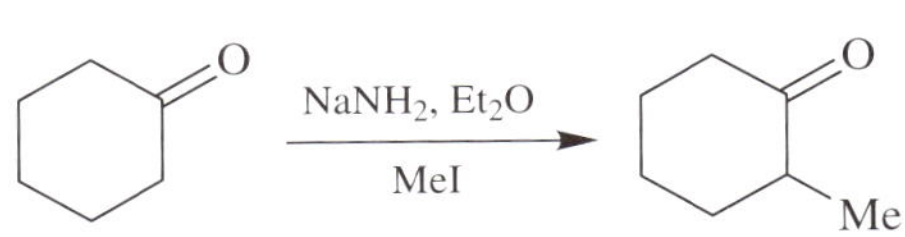

Scheme 3.78

One of the most important methods of C–C bond formation involves the **aldol condensation** in which the carbanion derived from an aldehyde or ketone reacts with a second carbonyl component to give a β-hydroxy-carbonyl derivative (Scheme 3.79).

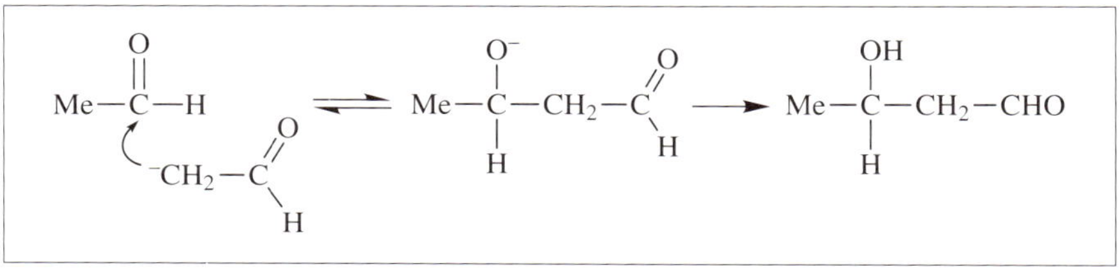

Scheme 3.79

Ketones are less reactive than aldehydes because the **inductive** and **steric effects** of the second alkyl group make the carbonyl group less susceptible to nucleophilic attack. Nevertheless, acetone can be induced to react and an equilibrium (see Scheme 3.80) can be established. Depending on the base, elimination of water may take place to give an unsaturated ketone. A second condensation may also take place to form phorone (2,6-dimethyl-4-oxohepta-2,5-diene).

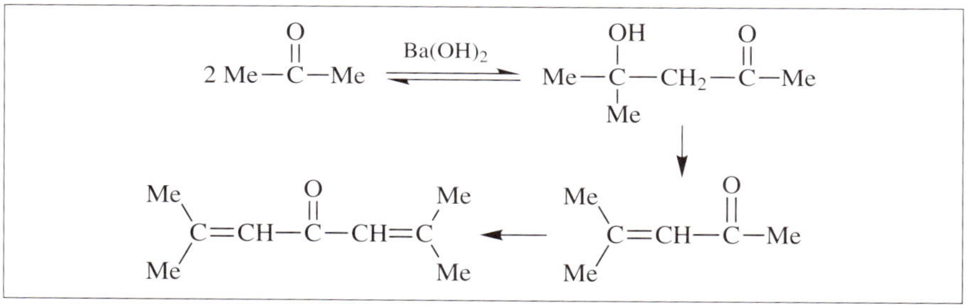

Scheme 3.80

The electron deficiency of the carbonyl group may be relayed through a double bond of an α,β-unsaturated ketone to the β-position. The addition of a carbon nucleophile to the β-position of an α,β-unsaturated ketone is known as a **Michael addition**. Since both the electron-withdrawing effect and the activating effect of the carbonyl group of phorone may be relayed through the separate double bonds, a further base-catalysed reaction can occur. This involves a Michael addition and leads to cyclization and the formation of isophorone (Scheme 3.81).

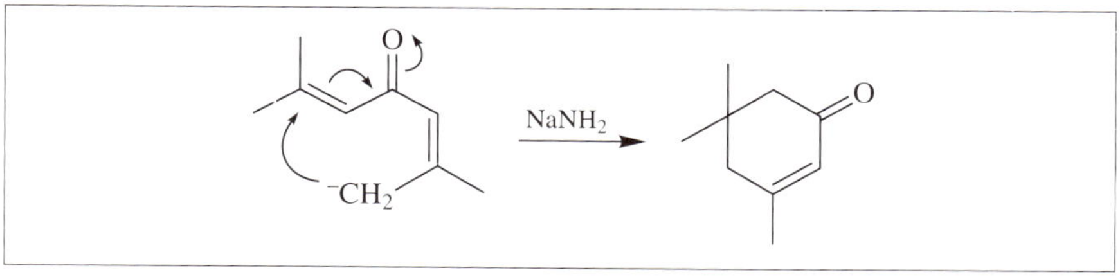

Scheme 3.81

If the carbonyl components are esters rather than ketones, somewhat stronger bases (for example, sodium ethoxide) are required to generate the carbanions. The product of the **Claisen condensation** between two molecules of an ester is a β-keto ester (Scheme 3.82).

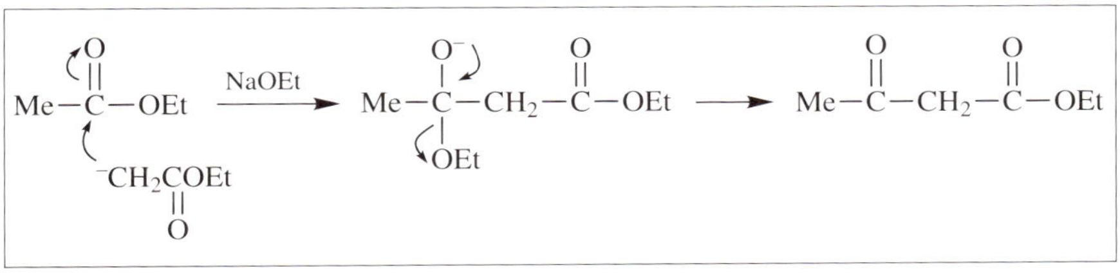

Scheme 3.82

The reaction can be used to form rings in the **Dieckmann cyclization** of a diester (Scheme 3.83).

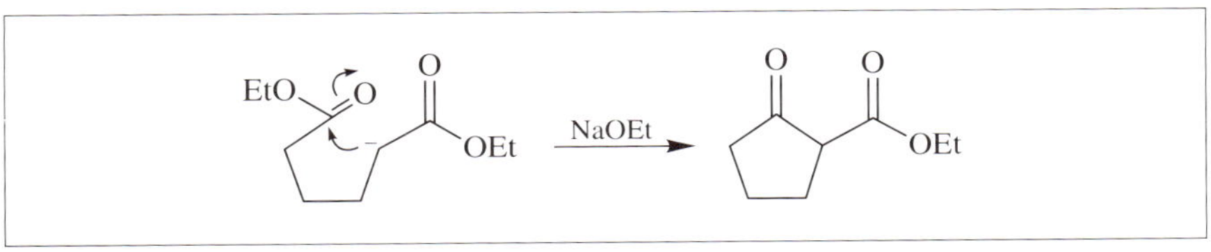

Scheme 3.83

There are a large number of variations on this general theme, in which both the source of the carbanion and the carbonyl recipient have been varied. The preparation of cinnamic acid and its ester from benzaldehyde exemplify some of these reactions (Scheme 3.84).

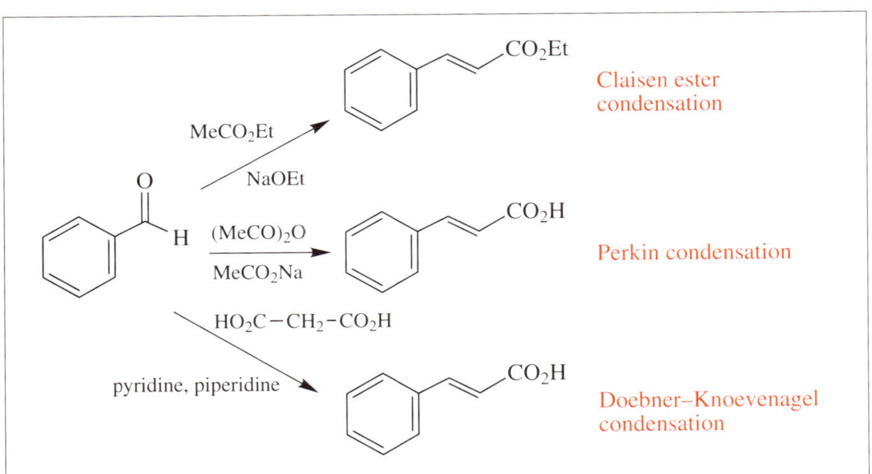

Scheme 3.84

The halogenation of positions adjacent to a carbonyl group takes place via the electrophilic attack of a halonium ion on the enol. Thus the bromination of ketones may be carried out with bromine in acetic acid.

Alternatively, the ketone may be converted to the enol acetate and this may be brominated (Scheme 3.85). Elimination of hydrogen bromide from α-bromo ketones leads to α,β-unsaturated ketones.

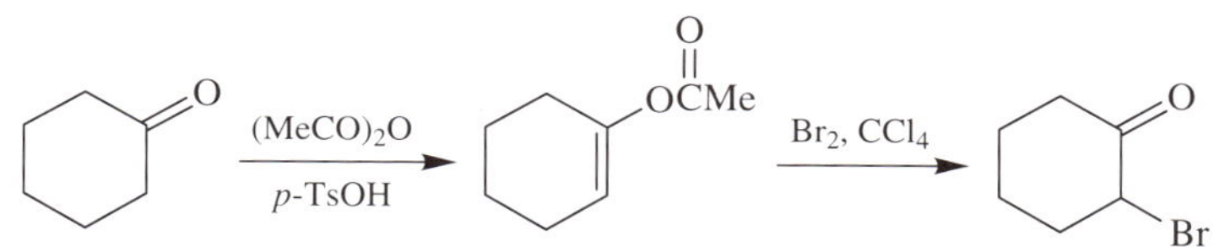

Scheme 3.85

A trihalo compound may be formed from the halogenation of a methyl ketone. The electron-withdrawing effect of the trihalomethyl group makes the carbonyl group very sensitive to nucleophilic addition. Consequently, in the presence of a mild base the trihalomethyl compound easily decomposes with the formation of chloroform, bromoform or iodoform, depending on the halogen. The other product of the haloform reaction is a carboxylic acid (Scheme 3.86).

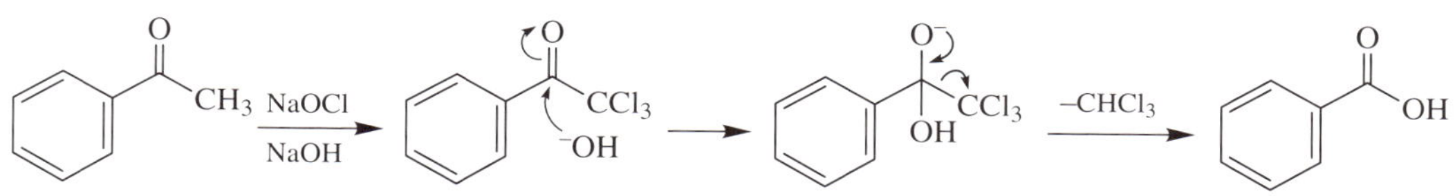

Scheme 3.86

The carbonyl group is not the only electron-deficient centre at which an enolate carbon may react. **Nitrosation** using nitrous acid provides a useful method of making C–N bonds.

The methylene group adjacent to a ketone may be oxidized by **selenium dioxide** to give 1,2-dicarbonyl compounds. It is important to carry out base-catalysed condensations of ketones wherever possible under an atmosphere of nitrogen. The reason is that enolate anions are readily **autoxidized** by the oxygen in air to form **hydroperoxides**. These may then undergo further reaction including decomposition to form diketones.

Exercise

Copy this revision chart showing the formation of other functional groups from enols and enolate anions, and fill in the relevant reagents and conditions for the inter-relationships beside the arrows. Note the consequences of using different sources for the carbanions.

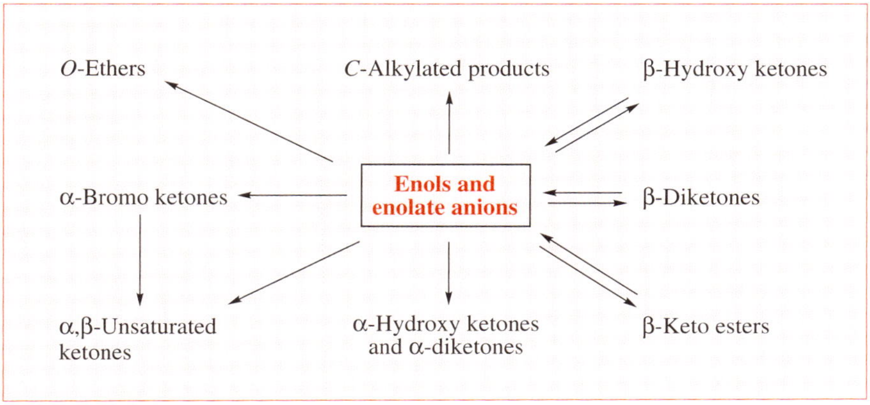

3.6 Nitriles, Imines and Nitro Compounds

In a nitrile (cyanide, RC≡N) or an imine (R_2C=N–) the nitrogen atom may be regarded as replacing the oxygen of a carbonyl group, whereas in a nitroso (RNO) or nitro (RNO_2) group the nitrogen may be regarded as replacing the carbon. Since nitrogen has an electronegativity between that of carbon and oxygen, it is not suprising to find similarities to carbonyl chemistry in both sets of compounds.

3.6.1 Nitriles

Alkyl nitriles show a large **dipole moment** consistent with polarization in the sense $RC^{\delta+}{\equiv}N^{\delta-}$. Hence many of the properties of the nitrile show a parallel to those of the carbonyl group. Nitriles are at the same oxidation level as a carboxylic acid, whereas an imine corresponds to an aldehyde or a ketone.

Nitriles may be prepared from amides by dehydration with phosphorus pentoxide, phosphorus oxychloride or thionyl chloride (Scheme 3.87).

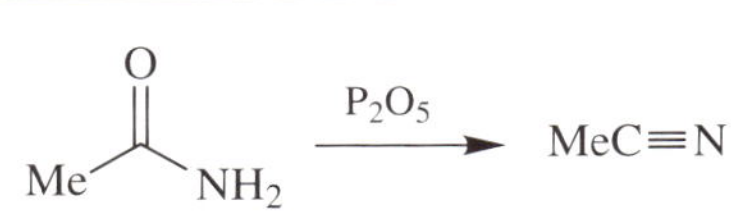

Scheme 3.87

An alternative route involves the nucleophilic substitution of an alkyl or benzyl halide with a cyanide ion. Two reactions are possible, leading to an isocyanide (R–N=C) or a cyanide (R–C≡N).

Many aryl nitriles are obtained from diazonium salts by treatment with copper(I) cyanide. Hydroxy and amino nitriles are obtained from addition reactions to carbonyl compounds.

Nitriles are reduced by sodium and alcohol or by the nucleophilic addition of a hydride from lithium aluminium hydride to give a primary amine (Scheme 3.88a). A partial reduction may be achieved by the use of tin(II) chloride and anhydrous hydrogen chloride. Hydrolysis at the end of the reaction releases the aldehyde. Free imines ($R_2C{=}NH$) are very rapidly hydrolysed to carbonyl compounds (Scheme 3.88b).

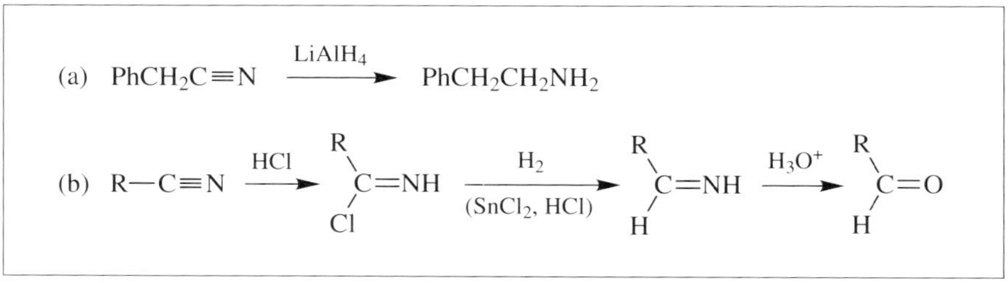

Scheme 3.88

Nitriles undergo hydrolysis under both acidic and alkaline conditions. Whereas alkaline hydrolysis normally gives the salt of the corresponding acid (Scheme 3.89a), acidic hydrolysis may be halted at the amide stage (Scheme 3.89b). The reaction may be combined with an esterification step if it is carried out with, for example, ethanolic sulfuric acid.

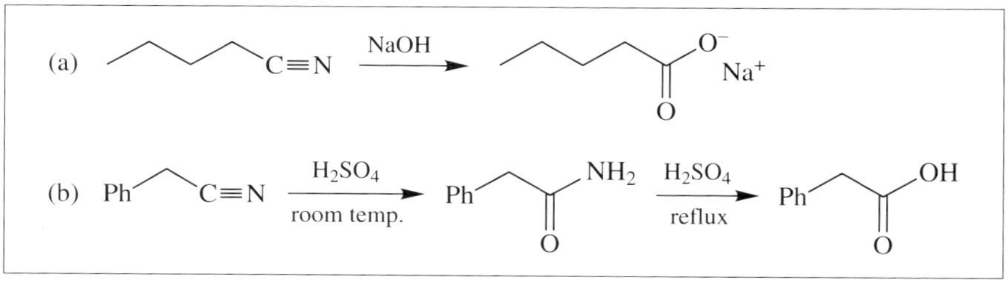

Scheme 3.89

If the nucleophile is the hydroperoxide anion derived from hydrogen peroxide, an imino hydroperoxide is formed. These are useful mild epoxidizing agents and decompose to give amides. Alkaline hydrogen peroxide can be used to hydrolyse nitriles.

The addition of carbon nucleophiles such as Grignard reagents gives, in the first instance, an iminomagnesium halide. This does not add a second alkyl residue because of the negative charge on the nitrogen. Hydrolysis of the imine at the end of the reaction releases the carbonyl group (Scheme 3.90). This provides a method of making ketones using a Grignard reagent.

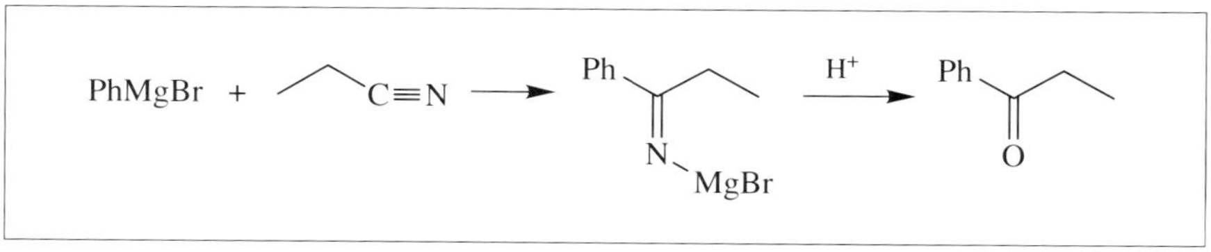

Scheme 3.90

Like the carbonyl group, the nitrile makes a hydrogen atom on an α-methylene weakly acidic. Treatment with a strong base ($NaNH_2$, $LiNEt_2$, NaH) generates a carbanion which can undergo alkylation and participate in various condensation reactions. The Thorpe–Ziegler method based on the cyclization of dinitriles provides a useful route for making cyclic ketones (Scheme 3.91).

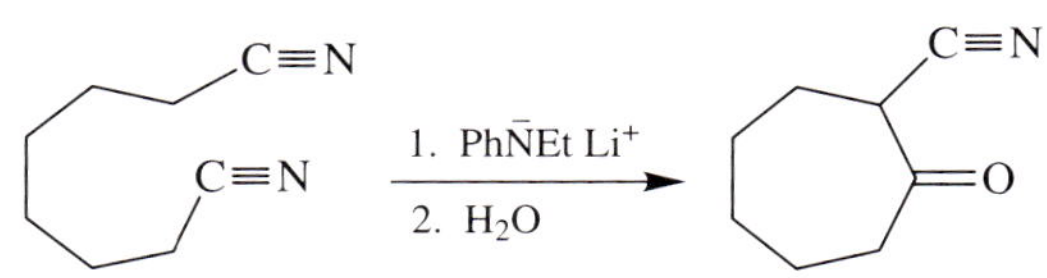

Scheme 3.91

The nitrogen atom of the nitrile possesses some nucleophilic characteristics. When a carbocation is generated by protonation of an alkene, it may be discharged by a nitrile. This reaction, known as the Ritter reaction, leads to the formation of amides (Scheme 3.92).

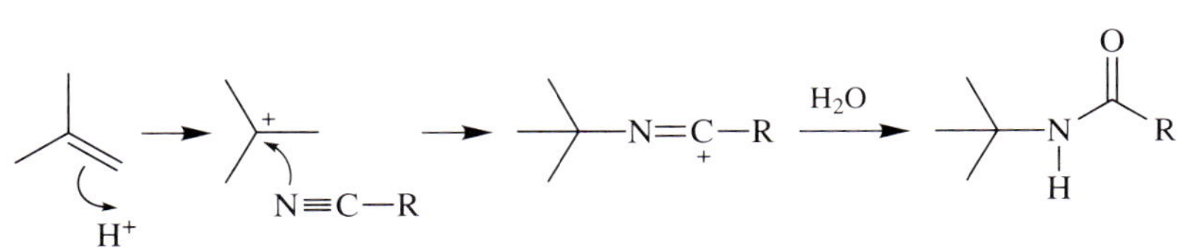

Scheme 3.92

3.6.2 Imines

Whereas isolated imines are relatively unstable and readily undergo hydrolysis to form carbonyl compounds, alkylated imines and imino derivatives are found as stable compounds in a variety of situations. Thus a number of nitrogen derivatives of aldehydes and ketones, such as the oxime, semicarbazone and hydrazone, contain an imino fragment.

Imines may be converted to their enamine tautomers, which contain an electron-rich alkene (Scheme 3.93).

Oximes are tautomers of *C*-nitroso compounds. In a number of instances the nitroso compound may be isolated as a dimer. Since nitroso

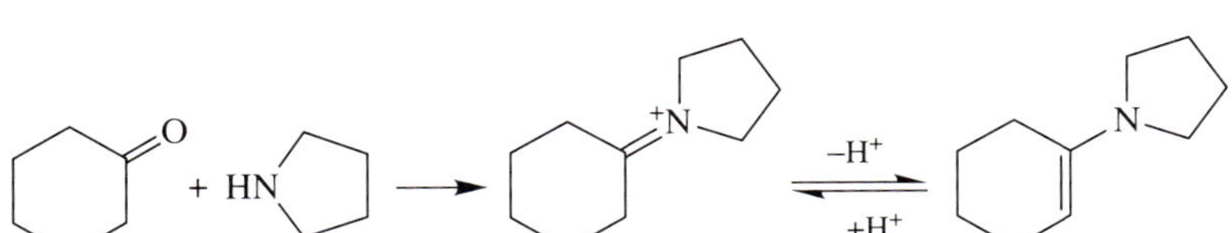

Scheme 3.93

compounds may be obtained from alkenes by the addition of nitrosyl chloride (NOCl), this tautomeric relationship allows a ketone to be introduced into a molecule.

3.6.3 Aliphatic Nitro Compounds

Aliphatic nitro compounds show a number of reactions which parallel those of carbonyl chemistry. Primary and secondary nitro compounds exhibit tautomerism paralleling keto–enol tautomerism (Scheme 3.94a). Aliphatic nitro compounds dissolve in aqueous sodium hydroxide with the formation of sodium salts. The resultant anions behave as carbanions and will condense with aldehydes. An example involves the formation of ω-nitrostyrene from nitromethane and benzaldehyde (Scheme 3.94b).

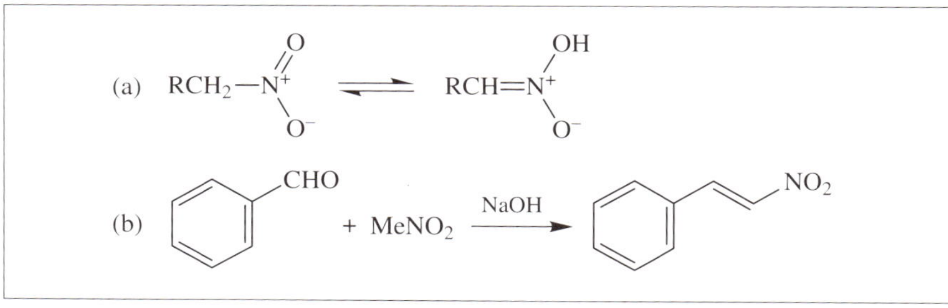

Scheme 3.94

Primary nitro compounds are hydrolysed by boiling hydrochloric acid to a carboxylic acid (Scheme 3.95a), but, under milder conditions, neutralization of a sodium salt with sulfuric acid leads to an aldehyde or ketone (Scheme 3.95b).

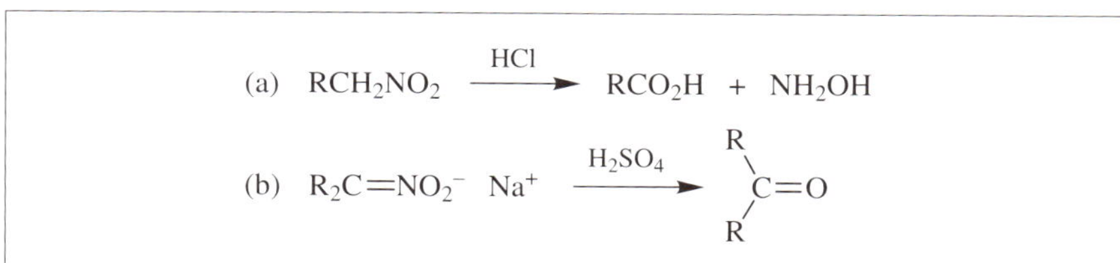

Scheme 3.95

Summary of Key Points

1. Alkenes may be prepared by elimination reactions with a regiochemistry (Hofmann or Saytzeff) that depends on the structure of the substrate and the reaction conditions. Alkenes may also be obtained from carbonyl compounds by the Wittig reaction and by the hydrogenation of alkynes.

2. Electrophilic addition to alkenes, such as the addition of hydrogen bromide under ionic conditions, follows the Markownikoff rule which states that in the addition of HX to an alkene, the hydrogen atom (the electrophile) becomes attached to the less-substituted carbon atom. The addition of halogens such as bromine proceeds *via* a halonium ion, and takes place with an overall *trans* stereochemistry.

3. *Syn* additions such as hydrogenation, epoxidation, hydroboration and osmylation involve *cis* addition from the less-hindered face of the alkene.

4. Pericyclic reactions such as the Diels–Alder reaction provide a useful method of creating new C–C bonds.

5. Alkynes undergo electrophilic, radical and nucleophilic addition, including a mercury-catalysed hydration to form methylene ketones.

6. The alkynyl C–H is sufficiently acidic to permit the formation of synthetically useful alkyne carbanions.

7. The chemistry of the carbonyl group is dominated by the electron deficiency of the carbon atom and its sensitivity to nucleophilic attack. Protonation of the carbonyl oxygen increases the electron deficiency and hence many reactions are carried out with acid catalysis.

8. There is a tautomeric relationship between carbonyl compounds and electron-rich enols which are susceptible to electrophilic attack.

9. Oxygen, sulfur, nitrogen and carbon nucleophiles add to the carbonyl group to form acetals, thioketals, imines and new C–C bonds.

10. Aldehydes may be distinguished from ketones by their ease of oxidation to acids.

11. The carbonyl group renders the hydrogen of an attached O–H, N–H or C–H acidic. The resultant anion is stabilized by resonance and acts as a nucleophile.

12. Tetrahedral intermediates are involved in the formation and reactions of esters and amides.

13. The acidity of the C–H lying between two carbonyl groups leads to synthetically useful carbanions.

14. The aldol condensation involves the reaction between a carbanion from an aldehyde or ketone and a second carbonyl component and leads to a β-hydroxy ketone. The Claisen condensation between two molecules of an ester leads to a β-keto ester.

15. The reactions of nitriles, imines, nitroso and nitro compounds show some parallels to carbonyl chemistry.

Worked Problems

Q Show how propyne (MeC≡CH) may be converted into: (a) $MeC{\equiv}CCO_2H$; (b) $MeCH{=}CHCMe_2OH$.

A The alkynyl hydrogen of propyne is acidic and hence the propyne will form an organometallic derivative. The Grignard derivative will react with carbon dioxide to give the carboxylic acid. *Cis* alkenes are formed by the catalytic hydrogenation of alkynes using a Lindlar catalyst. Hence a propynyl alcohol is a precursor to compound (b). This could be obtained by the addition of a propyne anion to propanone:

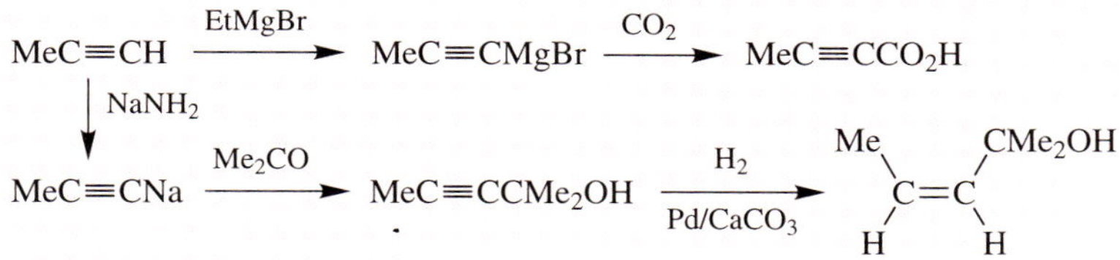

Q Suggest the most economical route for preparing $[1'\text{-}^{13}C]$epoxymethylenecyclohexane starting from $^{13}CH_3I$ and cyclohexanone.

A The epoxide can be prepared from cyclohexanone by reaction with the ylide derived from trimethylsulfoxonium iodide or in a two-step procedure via epoxidation of the methylenecyclohexanone prepared by a Wittig reaction. In preparing the trimethylsulfoxonium salt from [^{13}C]methyl iodide and dimethyl sulfoxide, the three methyl groups become equivalent and hence the label is diluted, whilst in the Wittig reaction using the ylide derived from methyltriphenyl phosphonium iodide, all of the labelled methyl group is transferred to the methylenecyclohexane. Hence the latter is the more economical:

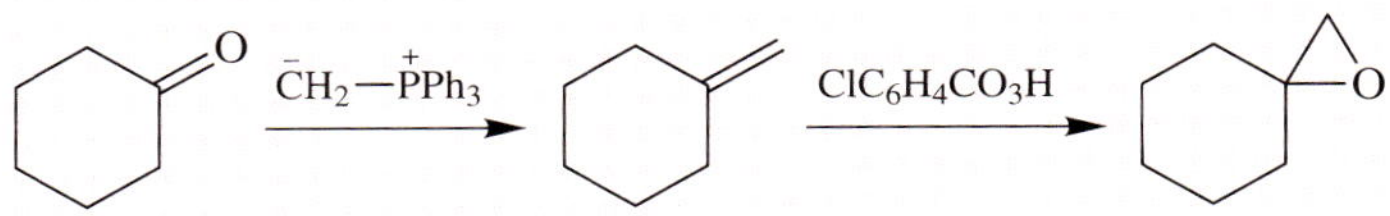

Q Give the structure of the product **A** of the reaction:

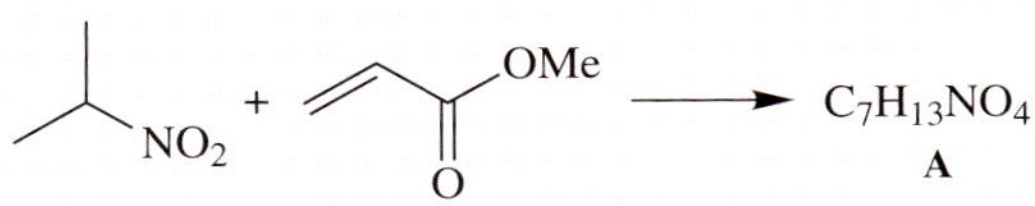

A The product **A** contains all the atoms of the precursor molecules and hence a simple addition has taken place. 2-Nitropropane has an acidic hydrogen and hence it can form a carbanion. Methyl acrylate is an unsaturated ester and can act as a Michael acceptor for a carbanion. The reaction may therefore be formulated as:

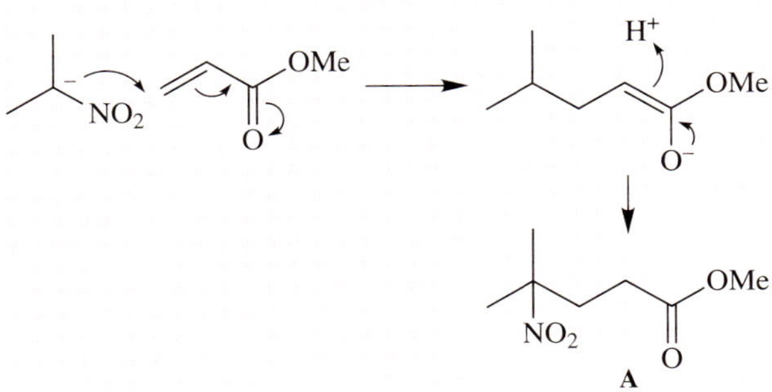

Problems

3.1. Give the product of the reaction of α-pinene (**1**) with the following reagents, showing the stereochemistry: (a) (i) BH_3•THF; (ii) H_2O_2, NaOH; (b) $ClC_6H_4CO_3H$; (c) SeO_2; (d) H_2/Pd; (e) OsO_4. (f) Treatment of the product of reaction (b) with acid gave sobrerol (**2**). Rationalize this reaction.

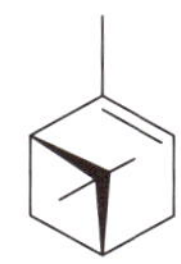

1

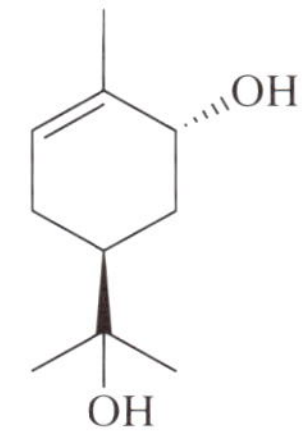

2

3.2. Show how propyne (MeC≡CH) may be converted into: (a) MeC(O)Me; (b) $MeCH_2CHO$; (c) $MeC{\equiv}CCH_2OH$.

3.3. Suggest ways of preparing the following deuteriated NMR solvents: (a) C^2HCl_3; (b) $(C^2H_3)_2CO$; (c) $C^2H_3CO_2{}^2H$.

3.4. Show how ethanal (acetaldehyde, MeCHO) may be converted into the following: (a) $MeCH_2OH$; (b) MeCOCl; (c) $MeCH_2NH_2$; (d) $MeC(O)CH_2Me$; (e) $MeCH{=}CHCO_2H$; (f) $MeCH(OH)CO_2H$.

3.5. Give the products of the following reactions:

(a) 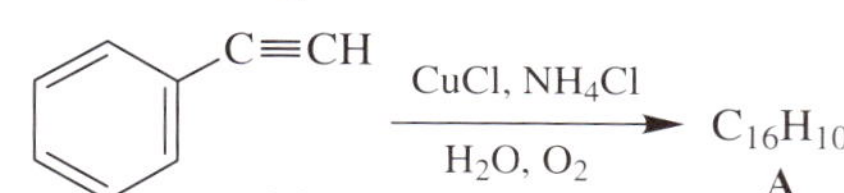

(b) 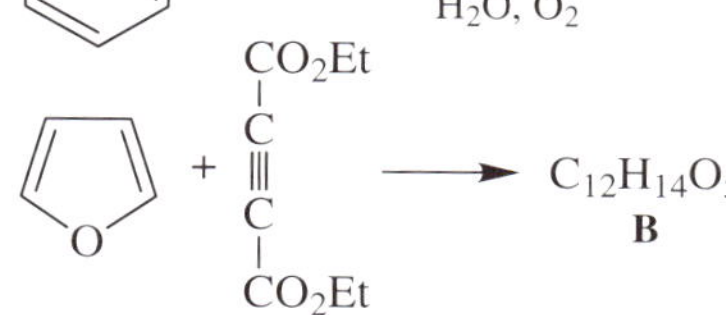

(c) 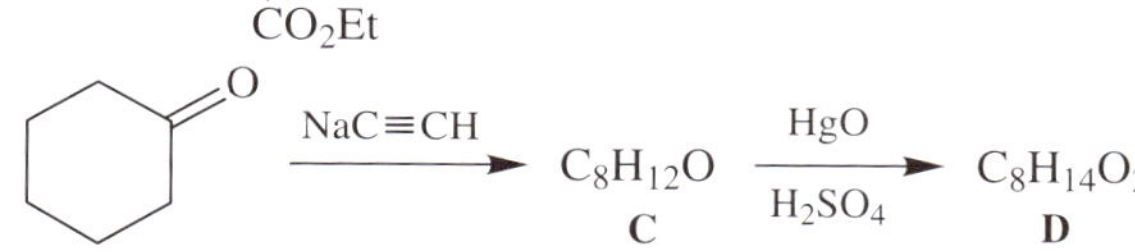

(d) 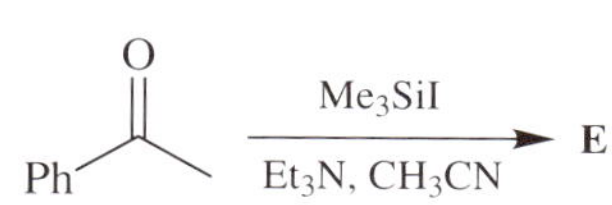

(e) 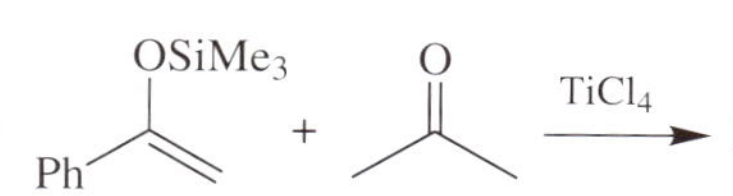

(f) 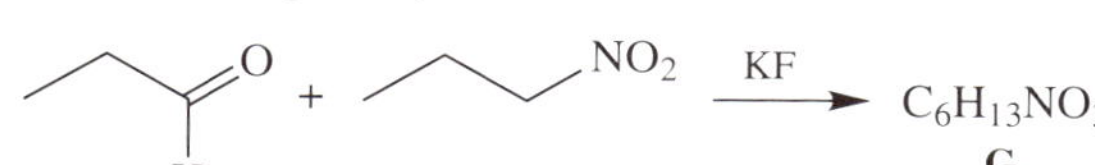

(g) 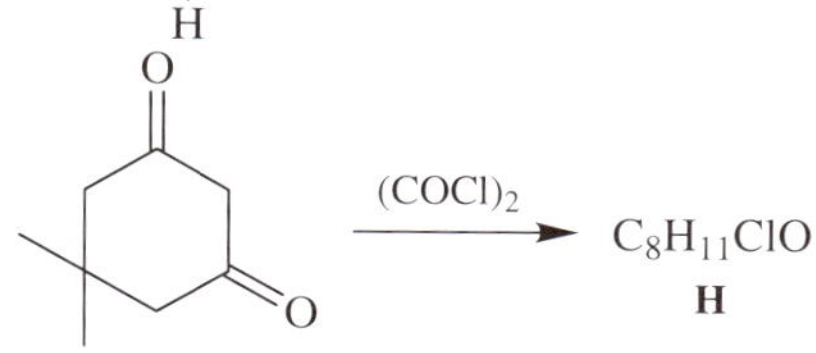

(h) 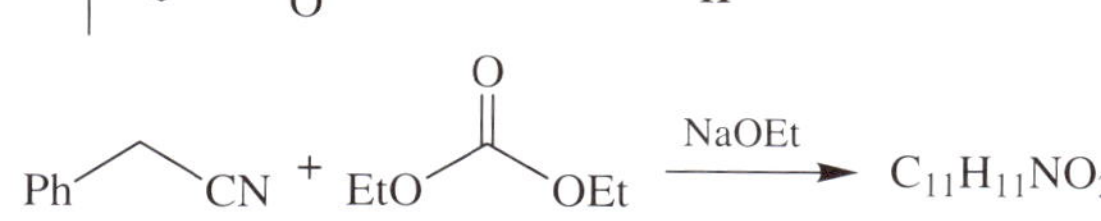

3.6. Indicate the missing reagents and reactants from the following sequence:

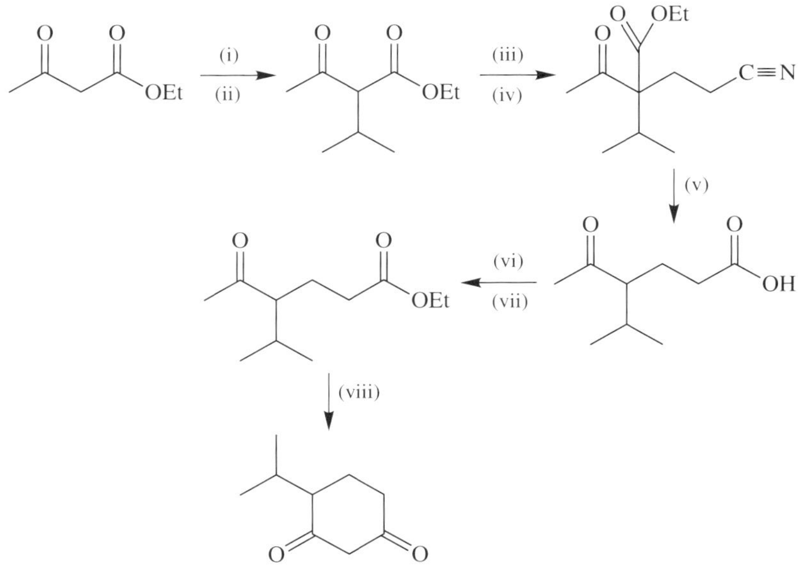

3.7. Provide a mechanism for the following reactions. (a) Treatment of the phosphonate **A** with formaldehyde in the presence of potassium carbonate gave the ester **B**.

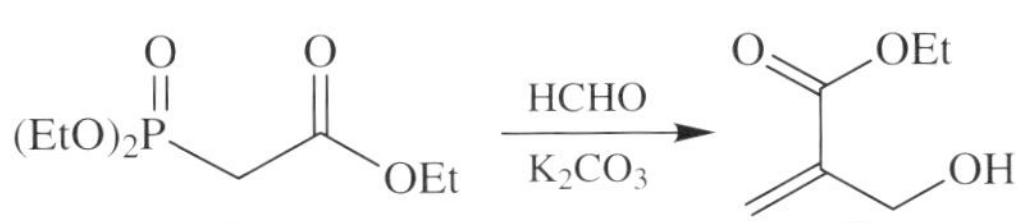

A **B**

(b) Treatment of 9-chloro-*trans*-1-decalone (**C**) with sodium methoxide in dimethoxyethane gave the methyl ester (**D**) after acidification.

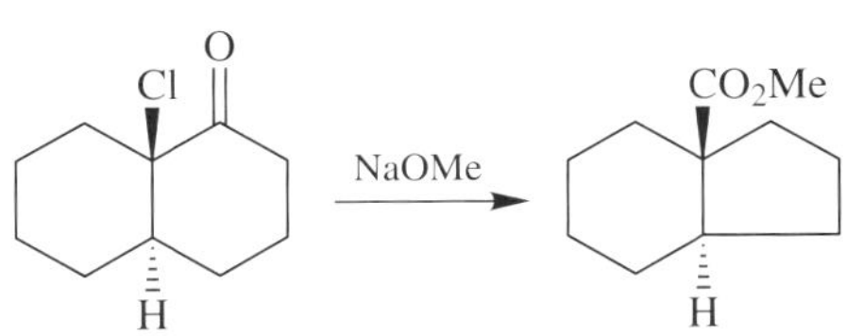

C **D**

(c) Treatment of *N*-benzyl-*N*-(hex-4-ynyl)amine (**E**) with formaldehyde in the presence of sodium iodide and acid gave the piperidine (**F**).

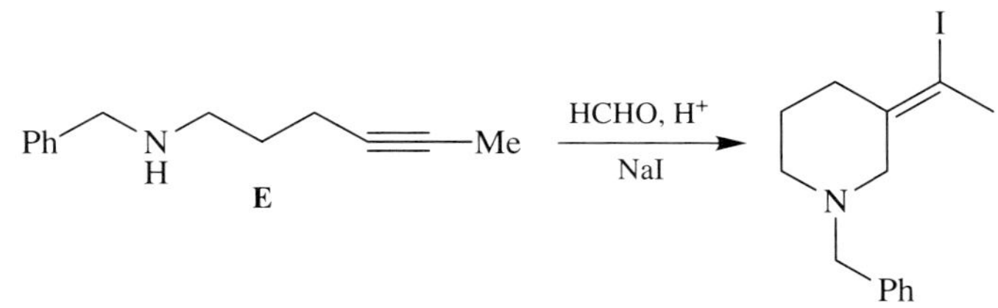

E → **F**

3.8. Treatment of β-chlorovinyl methyl ketone [ClCH=CHC(O)Me] with lithium acetylide (ethynyllithium) gave compound **A**, C_6H_7ClO. Reaction of compound **A** with dilute sulfuric acid gave compound **B**, C_6H_6O, which readily formed a 2,4-dinitrophenylhydrazone with 2,4-dinitrophenylhydrazine. Identify compounds **A** and **B**.

3.9. Treatment of ketene (CH_2=C=O) with HCN under pressure gave compound **A**, $C_5H_5NO_2$. This reacted with cyclopentadiene to give compound **B**, $C_{10}H_{11}NO_2$. Hydrolysis of compound **B** with base gave dehydronorcamphor (**3**). Identify compounds **A** and **B**.

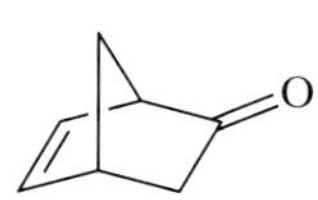

3

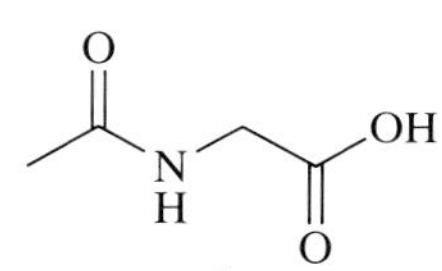

4

3.10. Treatment of *N*-acetylglycine (**4**) with benzaldehyde in the presence of sodium acetate and acetic anhydride gave compound **A**, $C_{11}H_9NO_2$, which did not possess OH or NH absorption in the infrared spectrum. When compound **A** was heated with water it gave acid **B**, $C_{11}H_{11}NO_3$. Identify compounds **A** and **B**.

3.11. Condensation of butyraldehyde (**5**) with diethyl malonate (**6**) in the presence of sodium ethoxide gave compound **A**, $C_{11}H_{18}O_4$. Treatment of **A** with 30% hydrogen peroxide in the presence of base gave compound **B**, $C_{11}H_{18}O_5$. Hydrolysis of compound **B** with alcoholic potassium hydroxide gave an acid which was heated in an oil bath to give compound **C**, $C_6H_{10}O_3$. This compound behaved as a monocarboxylic acid, giving a monomethyl ester. It also formed an oxime. Identify compounds **A**, **B** and **C**.

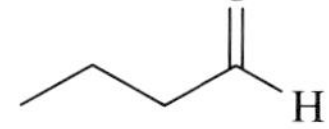

5

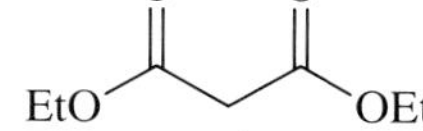

6

4
Chemistry of Aromatic Compounds

Aims

This chapter of the book describes the chemistry of the aromatic ring. At the end of this chapter you should be able to understand:

- That the presence of a cyclic conjugated system containing $(4n + 2)\pi$ electrons confers an additional stability on the molecule
- The electron-rich character of the aromatic system leads to reactions with electrophiles involving an addition–elimination pathway with the overall effect of substitution
- That electron-donating substituents, which increase the electron density of the aromatic ring, enhance the reactivity of the ring towards electrophiles, directing substitution to the *ortho* and *para* positions
- That electron-withdrawing substituents, which decrease the electron density of the aromatic ring, deactivate it towards electrophilic attack and direct substitution to the *meta* position
- That an electron-withdrawing group may facilitate nucleophilic substitution of halogen atoms in the *para* position by an addition–elimination mechanism
- That these effects may rationalize the reactivity of polycyclic aromatic compounds
- That the introduction of a heteroatom such as nitrogen into the ring system may lead to π-deficient (pyridine) and π-excessive (pyrrole) heteroaromatic compounds

4.1 Aromatic Substitution

The chemistry of aromatic systems is dominated by the enhanced stability associated with the presence of a cyclic conjugated system containing $(4n$

+ 2)π electrons. In the case of benzenoid aromatics where $n = 1$, this is a 6π system. For naphthalene (Box 4.1), $n = 2$ (10π electrons), and for anthracene and phenanthrene, $n = 3$ (14π electrons). A number of heteroaromatic rings (see Section 4.3) exist, such as pyridine, pyrrole, thiophene and furan, in which a carbon atom or atoms has been replaced by nitrogen, sulfur or oxygen. In the five-membered series the heteroatom contributes two electrons from its lone pair to the sextet.

Box 4.1 Some Aromatic Hydrocarbons

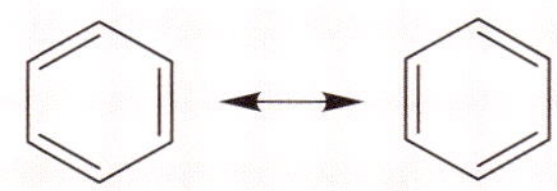

benzene

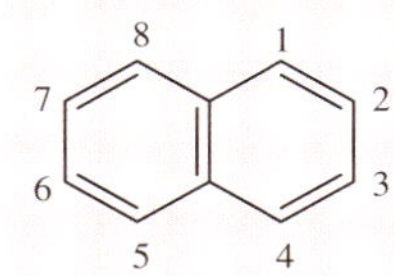

naphthalene

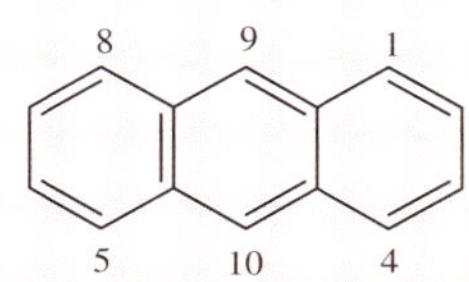

anthracene

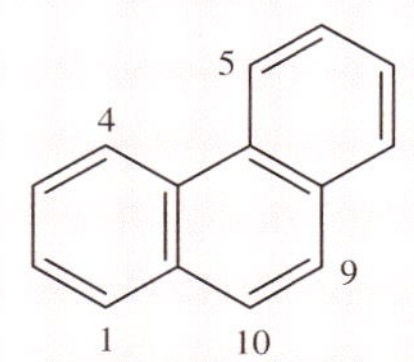

phenanthrene

The enhanced stability arising from the property of **aromaticity** is known as the **resonance energy** and for benzene this is 150.6 kJ mol^{-1}.

4.1.1 Electrophilic Aromatic Substitution

The reactions of the electron-rich π-system with electrophiles involve an addition followed by an elimination and reversion to an aromatic system. The picture of aromatic substitution which has emerged, and which is dealt with in more detail in books on reaction mechanisms, is shown in the reaction coordinate diagram (Figure 4.1) in which the **Wheland intermediate**, formed by the addition of the electrophile, represents a key step.

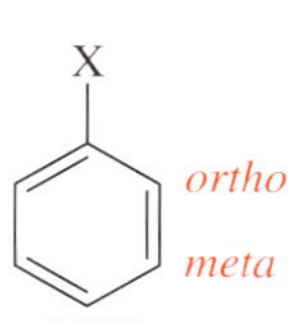

4.1

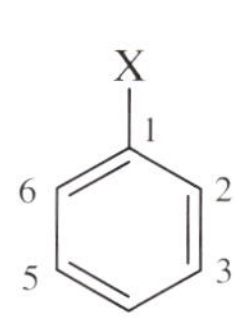

4.2

In considering the aromatic ring as a functional group, we have to examine the influence of substituents on the overall electron density of the aromatic ring, and their effect at specific sites (see **4.1** and **4.2**). Secondly, we have to consider the influence of these substituents on the Wheland intermediate.

There are other reactions of aromatic compounds, such as metallation reactions, which involve the C–H σ-system.

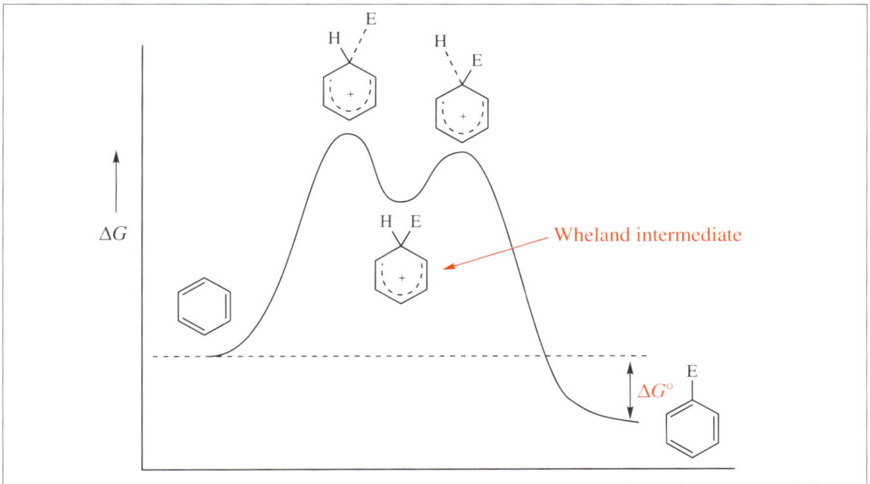

Figure 4.1

Substituent Effects in Electrophilic Substitution

Substituents fall into a number of groups. There are those which **activate** the ring by electron donation and can also stabilize the Wheland intermediate. Secondly, there are those which **deactivate** the ring by electron withdrawal, but their lone pairs can stabilize a Wheland intermediate. Finally, there are substituents which both deactivate the ring and destabilize the Wheland intermediate.

The activating influence of an electron-donating substituent is greatest at the 2- (*ortho*) and 4- (*para*) positions, whilst the deactivating effect of an electron-withdrawing substituent is least at the *meta* position. The orienting effects of substituents are summarized in Table 4.1.

Table 4.1 Directing effects in aromatic substitution

Ortho/para directing; activating	*Meta directing; deactivating*
R_2N, NH_2 (but salt formation can lead to deactivating NH_3^+), NHCOMe, OR, OH, alkenyl, aryl, alkyl, halides	CX_3 (X = halogen), CO_2H, C(O)R, SO_3H, NO_2, NR_3^+

The consequences of the nitration of a series of monosubstituted aromatic compounds using a nitric acid–sulfuric acid nitrating mixture are given in Table 4.2

In some cases the electrophile attacks the site bearing the substituent and the latter is then displaced. This is known as *ipso* substitution (Latin *ipse* = self). An example involves the nitration of isopropylbenzene in which propene is lost and nitrobenzene is formed.

Table 4.2 Nitration of monosubstituted aromatic compounds

Original substituent	*Orientation of nitro group* (%) *ortho*	*meta*	*para*
Me	58.5	4.4	36.8
OH	59.2	2.7	38.1
Cl	17.5	1.2	81.3
MeCONH	19.4	2.1	78.5
CO_2H	22.3	76.5	1.2
NO_2	8.1	90.9	1.0
SO_3H	19.9	68.3	6.2

The effect of a substituent on the reactivity of a particular centre may be quantified in terms of the partial rate factor. The partial rate factor is defined as the rate of substitution at a given position relative to that in any one position in benzene itself. Partial rate factors may be calculated by treating an equimolar mixture of benzene and the substituted benzene with insufficient reagent to complete the reaction. Analysis of the products will then show which substrate has reacted with more of the reagent and at which centre.

Although a detailed discussion of the influence of aromatic substituents on the rates of aromatic substitution is outside the scope of this book, it is nevertheless helpful to note that quantitative correlations of directing effects have been made using various forms of the Hammett equation:

$$\log(k/k_n) = \rho^+\sigma^+$$

where k and k_n are the rate constants for the substituted and unsubstituted compounds, respectively, whilst ρ^+ and σ^+ are the reaction and substituent constants. In the Hammett equation, which is used for *meta* and *para* substitution, the rate of reaction of a substituted benzene is compared to the rate for the unsubstituted benzene, and shown to be the product of a reaction constant (ρ^+) and a substituent constant (σ^+). The reaction constant reflects the susceptibility of the particular aromatic reaction to changes in the nature of the substituent, and the substituent constant measures the sum of the polar and resonance effects of a substituent on aromatic substitution. These are discussed in more detail in the book on reaction mechanisms in this series.

Reagents for Aromatic Electrophilic Substitution

The generation of an electrophilic chloronium or bromonium ion from chlorine or bromine, by heterolytic fission in an ionizing solvent such as glacial acetic acid, leads to a reagent that can be used for aromatic substitution (Scheme 4.1a), particularly of the more reactive ring systems. Indeed, under some circumstances, for example with phenol or aniline, polysubstitution may occur (Scheme 4.1b). A protic or a Lewis acid (H_2SO_4, $AlCl_3$, $FeBr_3$) catalyst may be used with less reactive species.

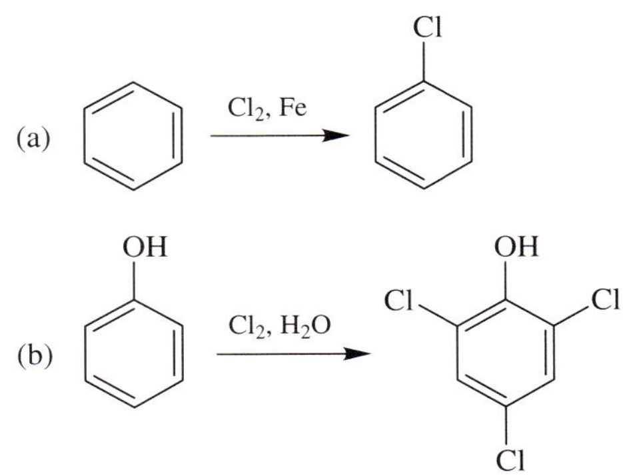

Scheme 4.1

For most iodinations the reactive species is not the iodonium ion. Iodinations require iodine and an oxidant. The actual reactant may be an iodosulfuric acid or iodonitric acid species.

The electrophilic nitronium ion (NO_2^+) can be generated from concentrated nitric acid by the action of concentrated sulfuric acid (the classical 'mixed acid'; Scheme 4.2) or with acetic acid and/or acetic anhydride. The latter may involve the formation of acetyl nitrate. Some metal nitrates, such as copper(II) nitrate or vanadium(V) oxytrinitrate, have been employed in nitrations, whilst the combination of ozone and nitrogen dioxide has provided a relatively non-acidic methodology, based on the formation of N_2O_5 and its fission to give $NO_2^+NO_3^-$.

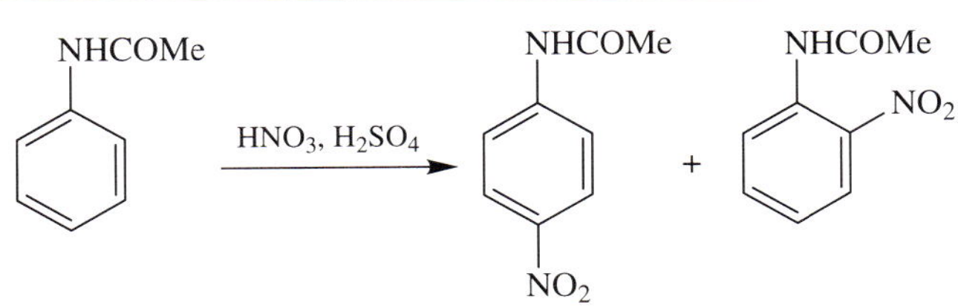

Scheme 4.2

A number of nitrations, particularly of phenols, may in practice be nitrosations followed by oxidation of the nitroso compound to form the nitro compound by nitric acid. Although the underlying pattern of the nitration (*ortho/para* or *meta*) is determined by the substituent(s) on the aromatic ring, the ratio of the different isomeric nitro compounds that are formed may vary with the reagent and the temperature at which the reaction is carried out.

The **diazonium salt** is another useful electrophilic nitrogen species, particularly with reactive aromatic rings such as phenols (Scheme 4.3). Further modification of the resultant coupled product may leads to amines or to hydrazines.

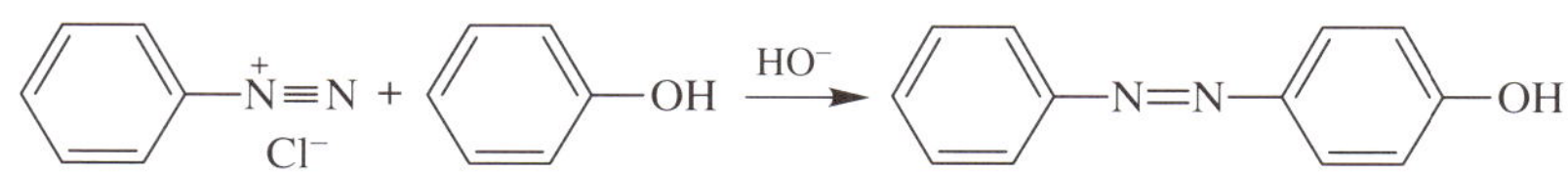

Scheme 4.3

Sulfonation or **chlorosulfonation** using fuming sulfuric acid or chlorosulfonic acid are useful reactions. Sulfonations are often reversible and hence the sulfonate group may be used to protect a particular site on an aromatic ring. It can then be removed later in a synthetic sequence.

The reaction with chlorosulfonic acid affords the sulfonyl chloride rather than the sulfonic acid (Scheme 4.4).

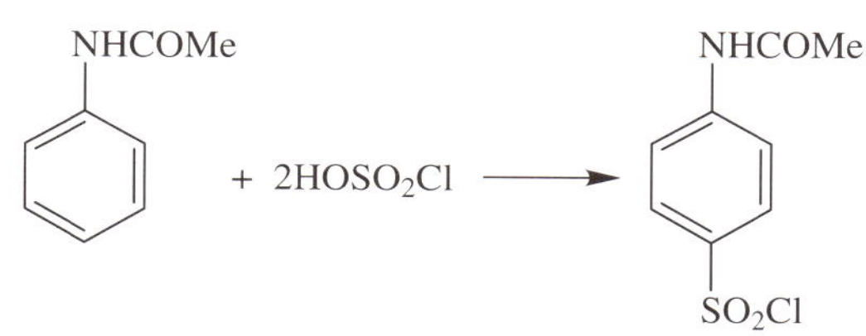

Scheme 4.4

Carbonium ions can be generated at a variety of oxidation levels. The **alkyl carbocation** can be generated from alkyl halides by reaction with a Lewis acid (RCl + $AlCl_3$) or by protonation of alcohols or alkenes. The reaction of an alkyl halide and aluminium trichloride with an aromatic ring is known as the **Friedel–Crafts alkylation**. The order of stability of a carbocation is tertiary > secondary > primary. Since many alkylation processes are slower than rearrangements, a secondary or tertiary carbocation may be formed before aromatic substitution occurs. Alkylation of benzene with 1-chloropropane in the presence of aluminium trichloride at 35 °C for 5 hours gave a 2:3 mixture of *n*- and isopropylbenzene (Scheme 4.5). Since the alkylbenzenes such as toluene and the xylenes (dimethylbenzenes) are more electron rich than benzene itself, it is difficult to prevent polysubstitution and consequently mixtures of polyalkylated benzenes may be obtained. On the other hand, nitro compounds are sufficiently deactivated for the reaction to be unsuccessful.

The lone pairs on an adjacent oxygen atom may stabilize a carbocation, affording a useful reactive species (Scheme 4.6a). The combina-

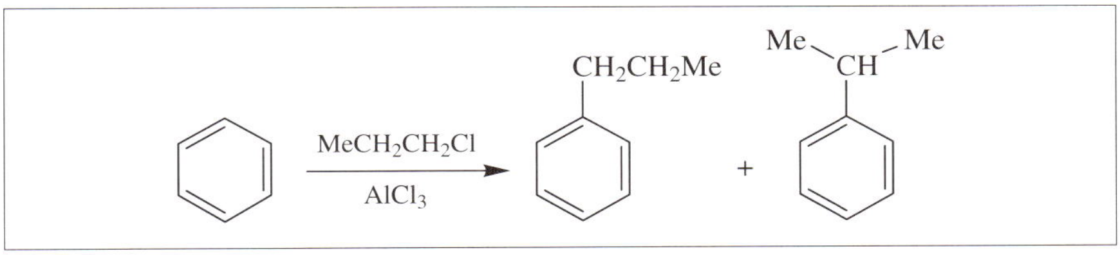

Scheme 4.5

tion of formaldehyde with zinc chloride and hydrochloric acid gives a reagent that may be used in the **chloromethylation** of aromatic hydrocarbons. Thus benzene affords chloromethylbenzene (Scheme 4.6b).

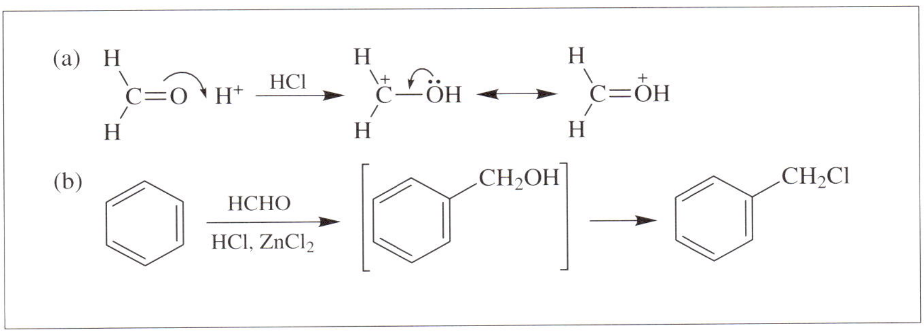

Scheme 4.6

The **acylium ion** (RCO^+), which may be generated from an acyl halide and a Lewis acid catalyst, also achieves some stabilization from oxygen lone pairs, and the carbocation remains localized on the carbonyl group. Consequently, Friedel–Crafts acylation is not accompanied by the rearrangements that affect alkylations (Scheme 4.7).

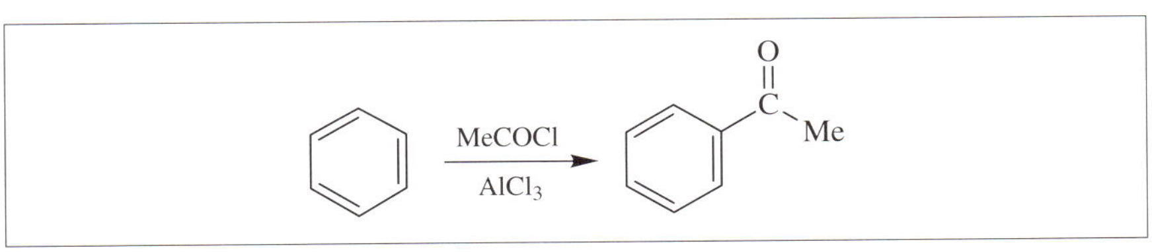

Scheme 4.7

Aromatic Substitution of Naphthalene

Naphthalene undergoes electrophilic aromatic substitution at C-1 more easily than at C-2. There is a smaller loss of resonance energy in forming the intermediate for reaction at C-1 and reaction takes place more rapidly at this centre. However, the products of aromatic substitution at C-1 suffer interactions with C-8 (*peri* interactions) and are less stable than the corresponding products of substitution at C-2. Hence those aromatic substitution reactions that are carried out under conditions that allow equilibration between isomers (thermodynamic control) lead to substitution at C-2, but reactions that are carried out under conditions that are essentially irreversible (kinetic control) lead to substitution at C-1. Sulfonation at 40 °C under kinetic control yields naphthalene-1-sulfonic acid, whilst at 160 °C equilibration takes place and the more stable naphthalene-2-sulfonic acid is formed.

The directing effects of substituents can be summarized as follows. Activating groups such as hydroxyl, methoxyl, amino and amide, alkyl and halogen in the 1-position direct the electrophilic substitution to the 4-position, with some 2-substitution. If these groups are at the 2-position, then attack is directed to the 1-position. Deactivating groups such as nitro or carboxyl direct the substitution to the other ring and to positions 5 and 8.

Nitration using nitric acid gives 1-nitronaphthalene and then 1,5- and 1,8-dinitronaphthalenes. Bromination with bromine in carbon tetrachloride gives 1-bromonaphthalene followed by 1,4-dibromonaphthalene, whilst chlorination with sulfuryl chloride (SO_2Cl_2) affords 1-chloronaphthalene. Friedel–Crafts reactions can proceed with thermodynamic control and give mixtures of 1- and 2-acetyl and 1- and 2-methyl derivatives. The higher homologues [ethyl bromide with iron(III) chloride] give 2-substitution. Naphthalene is readily oxidized to 1,4-naphthaquinone by chromium(VI) oxide, whilst catalytic hydrogenation takes place in a stepwise manner to give tetralin and then decalin.

4.1.2 Aromatic Nucleophilic Substitution

Hitherto we have concentrated on electrophilic aromatic substitution. However, certain π-deficient aromatic rings are deactivated towards electrophilic attack but are susceptible to nucleophilic addition and a subsequent elimination. A particular example is 2,4-dinitrochlorobenzene. The electron-withdrawing nitro groups facilitate a Michael-type addition of a nucleophile to give a so-called Meisenheimer intermediate (Scheme 4.8). Collapse of the Meisenheimer intermediate and reversion to the aromatic system may lead to expulsion of the halide ion, as exemplified by the preparation of 2,4-dinitrophenylhydrazine. 2,4-Dinitrofluorobenzene is known as Sanger's reagent and is used for the detection of the N-terminal amino acids in peptides.

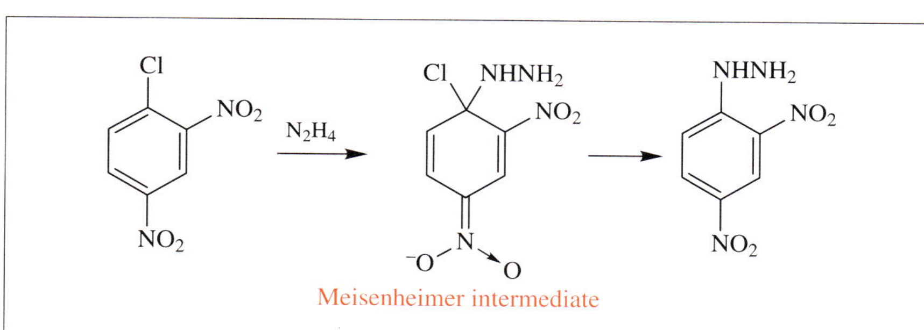

Scheme 4.8

4.1.3 Benzyne Intermediates

When chlorobenzene is heated with sodamide, aniline is formed. However, a direct replacement of the halogen does not occur. The reaction proceeds through a highly reactive **benzyne** formed by elimination in which the cleavage of the C–H bond is the rate-determining step. The amide anion may then attack the symmetrical benzyne at either end of the benzyne triple bond (Scheme 4.9).

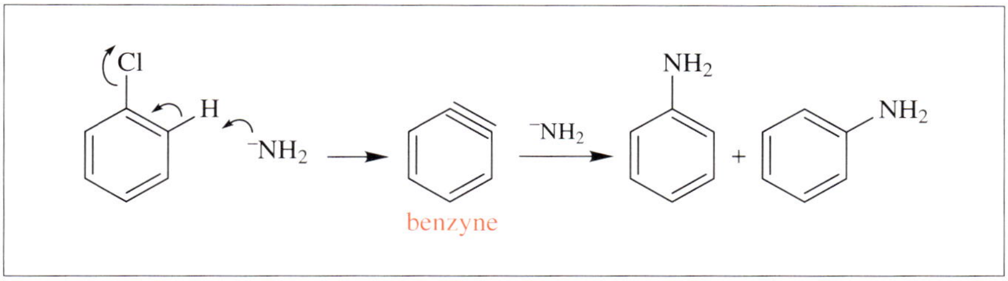

Scheme 4.9

Benzyne is a good **dienophile**. For example, reaction of 2-bromofluorobenzene with lithium metal or decomposition of benzenediazonium-2-carboxylate both generate benzyne, which can be trapped with furan to give a naphthalene endoxide (Scheme 4.10).

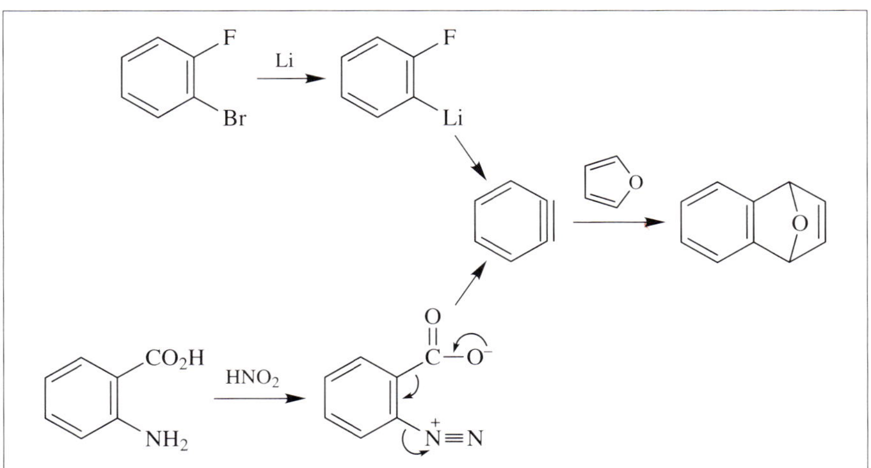

Scheme 4.10

4.2 Aromatic Functional Groups

4.2.1 Aromatic Halides

Aromatic halides, in which the halogen is attached to the ring, may be prepared by direct electrophilic aromatic substitution or from arylamines via diazonium compounds (see below).

Benzyl halides may be prepared by the free radical halogenation of the corresponding hydrocarbon using reagents such as *N*-chloro- or *N*-bromosuccinimide in the presence of light or a radical initiator. Other methods involve the substitution of benzyl alcohols. Benzyl chlorides may also be prepared by the chloromethylation of aromatic compounds.

An aromatic ring has a significant effect on the reactivity of the carbon–halogen bond. There are two very different systems: aryl halides, exemplified by chlorobenzene, and benzyl halides, exemplified by benzyl chloride. These illustrate the different effects that the presence of an aromatic ring can have on the reactivity of a functional group.

An aryl halide such as chlorobenzene is relatively unreactive towards nucleophilic substitution. The S_N1 and S_N2 pathways involve mechanisms that are not open to aryl halides. The greater 's' character of the sp^2 bond makes it more difficult to cleave the bond to generate a carbocation. However, these restrictions do not apply to radical or carbanion chemistry. Hence, aryl halides undergo radical coupling reactions and metal insertion reactions, leading to organometallic compounds.

These differences in reactivity are illustrated by the following. Chlorobenzene is unreactive on prolonged boiling with aqueous alkali, aqueous ammonia or aqueous potassium cyanide, conditions which would bring about replacement of the halogen in an alkyl halide. Heating with aqueous sodium hydroxide at 300–370 °C under pressure or passing chlorobenzene and steam over a silica catalyst at 400 °C are the type of conditions that are required to form phenol. When chlorobenzene is heated with aqueous ammonia and a small amount of a copper(II) sulfate catalyst at 200 °C, aniline is formed, together with small amounts of diphenylamine and phenol. These reactions probably occur *via* a benzyne intermediate. This lack of reactivity contrasts with the ease of nucleophilic substitution of benzyl halides by alkali and by amines

Although chlorobenzene does not easily give a Grignard derivative, bromobenzene reacts normally in diethyl ether with magnesium to give phenylmagnesium bromide. This undergoes typical Grignard reactions, for example with carbon dioxide to give benzoic acid (Scheme 4.11).

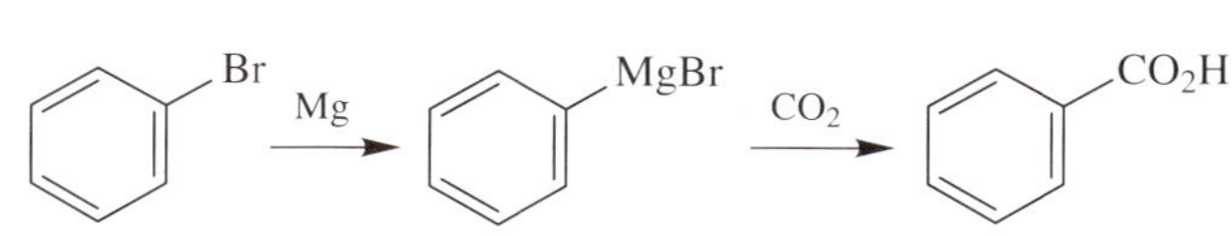

Scheme 4.11

Aryl bromides and iodides undergo a series of aryl coupling reactions mediated by organometallic derivatives. An example is the Heck reaction, in which an organopalladium iodide complex is formed and coupled with an alkene (Scheme 4.12).

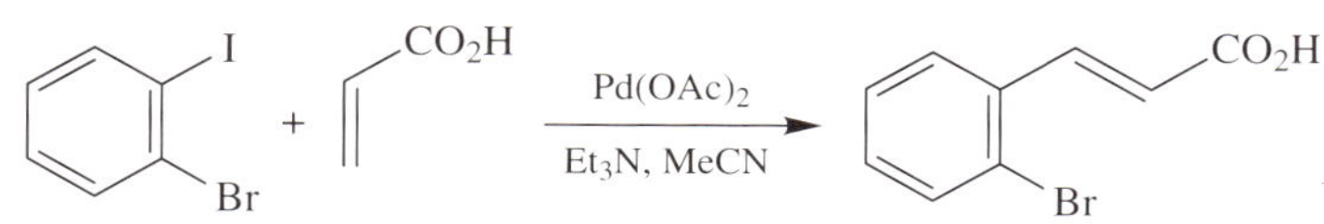

Scheme 4.12

The electrophilic aromatic substitution of aryl halides takes place less readily than with benzene (electron-withdrawing effect), but occurs at the *ortho* and *para* positions (the lone pairs on the halogen assist in delocalizing the positive charge in the intermediate). Further chlorination of chlorobenzene, in the presence of aluminium or iron trichlorides, gives 1,4-dichlorobenzene and some 1,2-dichlorobenzene. Nitration normally occurs to give the 2- and 4-nitro- and 2,4-dinitrochlorobenzenes (Scheme 4.13).

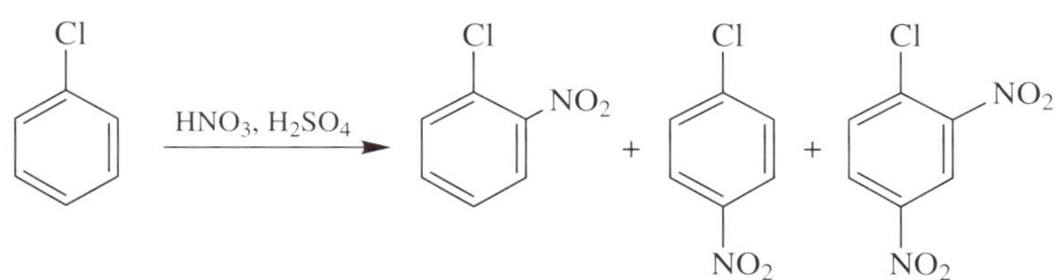

Scheme 4.13

Chloromethylation and the Friedel–Crafts alkylation and acylation take place normally at the *ortho* and *para* positions. The insecticide DDT (**4.3**, dichlorodiphenyltrichloroethane) is prepared from chlorobenzene and chloral (trichloroethanal).

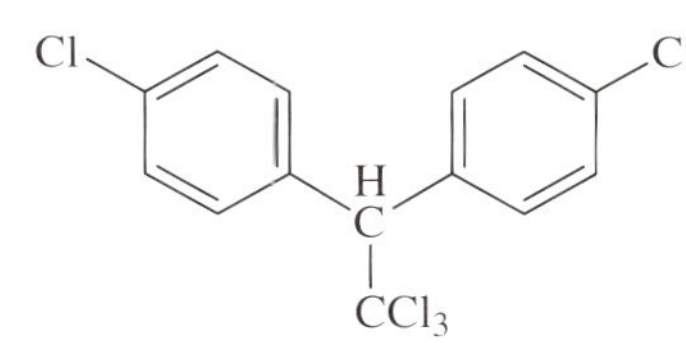

4.3 DDT

The presence of additional substituents on the aromatic ring may modify the reactivity of the halogen by opening up alternative reaction pathways. Thus an *ortho* or *para* nitro group makes an aryl halogen more susceptible to nucleophilic aromatic substitution.

4.2.2 Phenols

Phenols may be obtained by fusing the corresponding sulfonic acid with sodium or potassium hydroxide or by the hydrolysis of diazonium salts (see below). Both 1- and 2-naphthol may be obtained from the corresponding sulfonic acids.

On an industrial scale, phenol is obtained by the oxidation of isopropylbenzene (cumene). Initially a hydroperoxide is formed, which then undergoes a fragmentation and rearrangement. The initial oxidation illustrates the susceptibility of benzylic positions to oxidative, particularly radical, attack (Scheme 4.14).

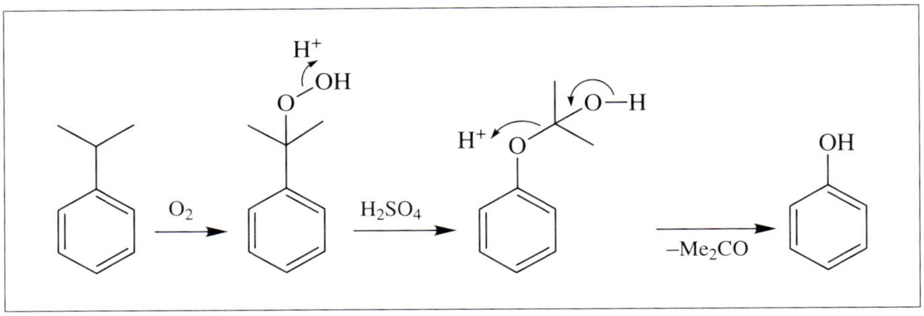

Scheme 4.14

The oxidative rearrangement of various acylbenzenes with alkaline hydrogen peroxide (the Dakin reaction) can provide a mild method for the synthesis of phenols (Scheme 4.15).

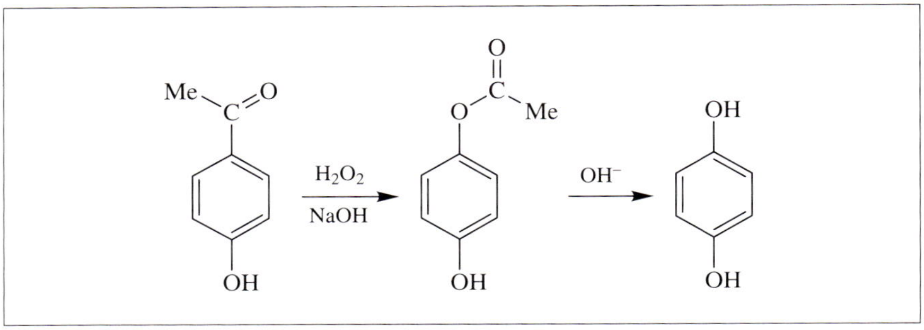

Scheme 4.15

The hydroxyl group of a phenol activates the aromatic ring towards electrophilic attack, whilst the aromatic ring increases the acidity of the hydroxyl group compared to an aliphatic alcohol. Thus many phenols are soluble in sodium hydroxide solution to form the phenoxide anion. Electron-withdrawing nitro groups in the *ortho* and *para* positions provide additional resonance stabilization for the phenoxide anion. 2,4,6-Trinitrophenol (picric acid) is quite a strong acid.

Like alcohols, phenols form ethers such as anisole (PhOMe) and esters such as phenyl acetate and phenyl benzoate ($PhCO_2Ph$).

A number of the reactions of phenols are typical of enols. Thus they are readily halogenated in the *ortho* and *para* positions, with a marked tendency for polyhalogenation. Bromination of phenol with bromine in carbon tetrachloride gives 4-bromophenol, but 2,4,6-tribromophenol is formed with bromine water.

Nitration takes place easily and again there is a tendency for polysubstitution. Phenols are readily nitrosated by nitrous acid. The resultant nitrosophenols are tautomeric with the corresponding quinone monoximes (Scheme 4.16).

Phenols readily form new C–C bonds under a variety of conditions, again paralleling the reactions of enols and enolate anions. For example,

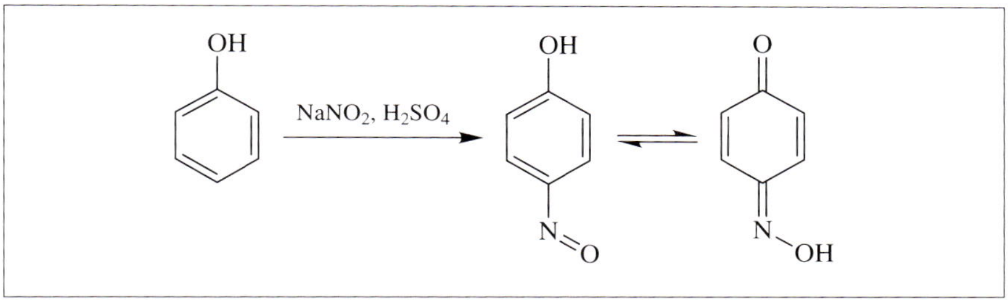

Scheme 4.16

phenols readily undergo *C*-acylation in the presence of quite mild acid catalysis. Phenol esters may undergo rearrangement in the presence of a Lewis acid (the Fries rearrangement), leading to *C*-acylation (Scheme 4.17).

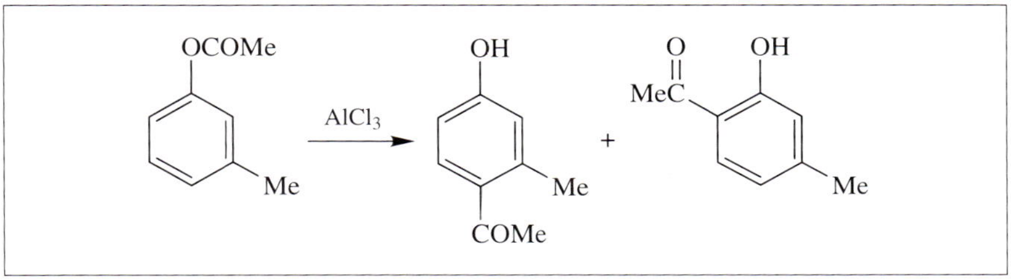

Scheme 4.17

Allyl phenyl ethers undergo an intramolecular [3,3]-sigmatropic rearrangement (the Claisen rearrangement) to form the *C*-alkyl derivative (Scheme 4.18). A consequence of the electrocyclic mechanisms is that the γ-carbon atom of the allyl ether becomes attached to the aromatic ring.

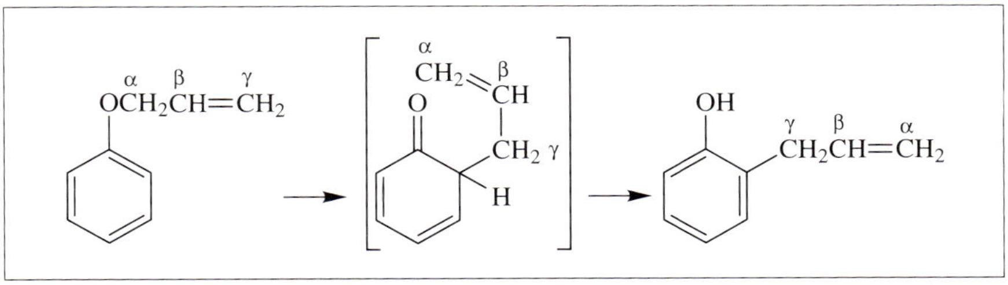

Scheme 4.18

The formation of the phenoxide anion enhances the reactivity of the *ortho* and *para* positions of the aromatic ring towards electrophilic reagents. The reaction of the phenoxide anion with carbon dioxide at 130 °C leads to *ortho* carboxylation (the Kolbe reaction). Thus phenol gives salicylic acid (**4.4**), the acetate of which is aspirin. The reaction is reversible and *ortho* phenolic acids undergo decarboxylation on heating.

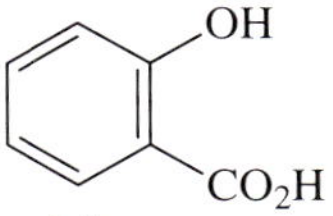

4.4

Phenols react easily with chloroform in the presence of base in the Riemer–Tiemann reaction. The reaction of chloroform with sodium hydroxide leads to the formation of dichlorocarbene (:CCl_2), and it is this which reacts with phenol to give the aldehyde (Scheme 4.19).

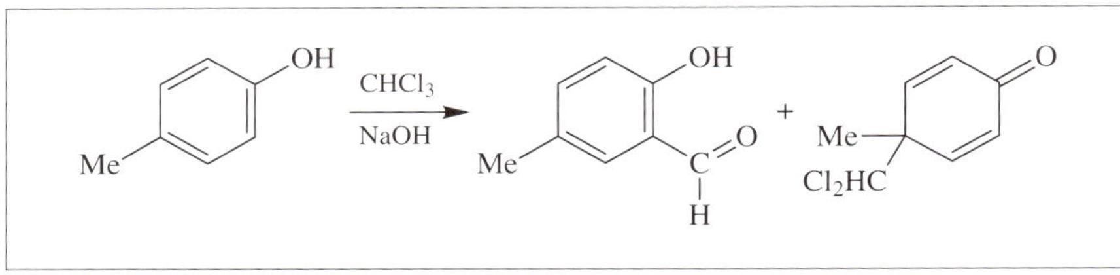

Scheme 4.19

Phenols can condense with aqueous formaldehyde in mildly alkaline solution with the introduction of a hydroxymethyl group in the *ortho* or *para* position to the phenolic hydroxyl group. In the presence of acid, phenols react with formaldehyde to give the phenol–formaldehyde resins such as Bakelite.

The coupling reactions of phenols with diazonium salts take place most readily using the phenoxide anions. Many of the products are dyestuffs (see below).

Phenols are quite sensitive to oxidation. On the one hand, they are easily oxidized to quinols and on further oxidation with, for example, iron(III) chloride, chromic acid or silver(I) oxide give *p*-quinones. However, under one-electron transfer conditions the phenoxide anion is oxidized to the phenoxyl radical. This shows free radical reactivity on the oxygen atom and at the *ortho* and *para* positions (Scheme 4.20a). The phenoxyl radical may readily dimerize. This is exemplified by the formation of **Pummerer's ketone** from *p*-cresol (Scheme 4.20b).

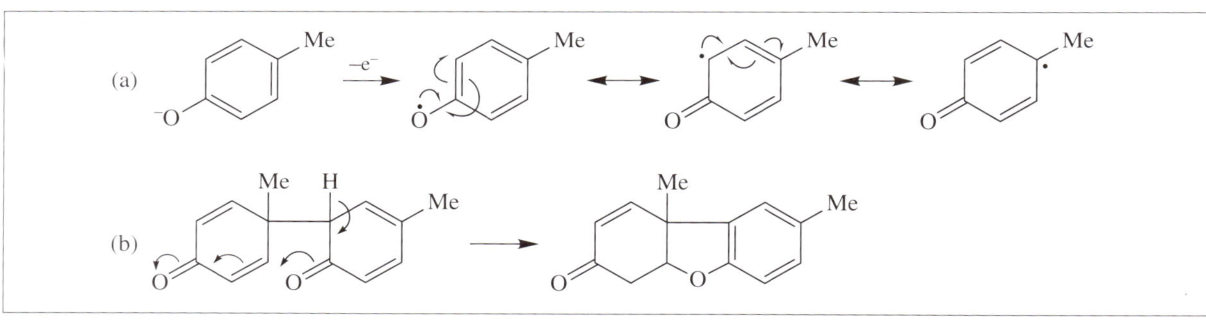

Scheme 4.20

The loss of resonance energy involved in the reduction of an aromatic ring makes this a difficult process. If the reaction is carried out using hydrogen and a catalyst, high pressure and temperature are required. However, reduction can be achieved by the stepwise addition of electrons using sodium or lithium in liquid ammonia, followed by protonation from an alcohol. This reaction, which has had wide application in the laboratory, is known as the **Birch reduction**. It is commonly used with phenol ethers and leads to the formation of 1,4-dihydroarenes. The resultant enol ethers may be hydrolysed by acid to give the corresponding unsaturated ketones (Scheme 4.21).

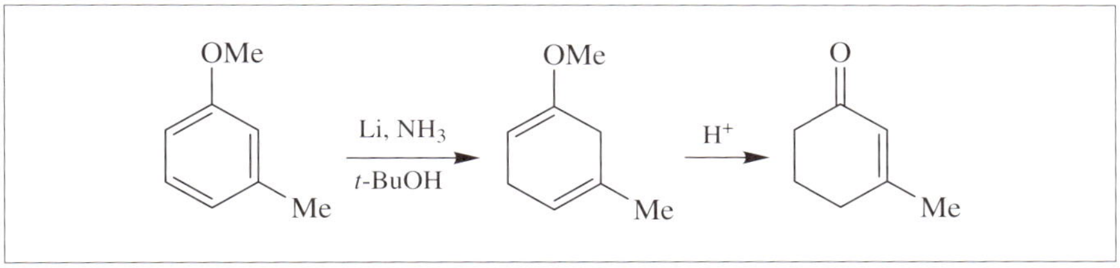

Scheme 4.21

4.2.3 Aromatic Amines

Aromatic amines are prepared by the reduction of nitro compounds under acidic conditions; this leads to the protection of the resultant amine as the salt. The selective reduction of one nitro group in 1,3-dinitrobenzene can be achieved with sodium polysulfide.

Aromatic amines are less basic than their aliphatic counterparts. The lone pair on the nitrogen atom is delocalized over the π-system of the aromatic ring. Furthermore, the basicity is influenced by substituents on the aromatic ring. Thus 4-nitroaniline (**4.5**) is less basic than aniline. However, the aromatic amines behave as typical nucleophiles. Alkylation of the nitrogen by alkyl halides leads to the mono- and dialkylanilines. Reaction with acetic anhydride or acetyl chloride affords acetanilides.

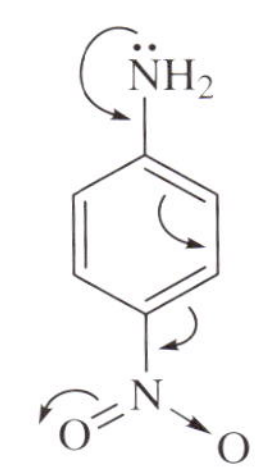

4.5

Aromatic amines, such as aniline, are easily chlorinated with chlorine water to give, for example, 2,4,6-trichloroaniline.

The reactivity of the aniline may be modified by acetylation. Bromination of acetanilide with bromine in acetic acid gives 4-bromoacetanilide (**4.6**). Under some conditions, bromination can take place first on the amide to give the *N*-bromo compound. This then undergoes a rearrangement known as the **Orton rearrangement** to give the *para* substituted product.

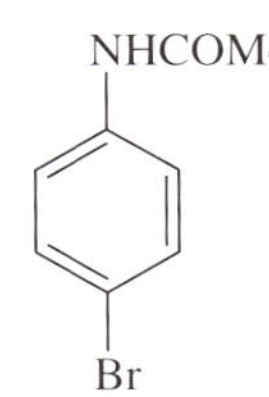

4.6

Although protonation of the amine leads to the **anilinium salt**, which is a deactivating substituent, nitration of aniline in nitric acid–sulfuric acid yields 60% *para* and 34% *meta* substitution, whilst nitration of acetanilide gives 79% *para* and 19% *ortho* nitroacetanilides.

Although carbon electrophiles can be used to substitute the aromatic ring, these reactions are difficult to achieve because of the tendency of the reagents to react on the nucleophilic nitrogen.

Oxidation of the aromatic amines occurs readily to give the aniline blacks. These reactions are of historic interest. The oxidation may take place *via* the quinone (**4.7**) or quinone imine, which subsequently react with unchanged amine.

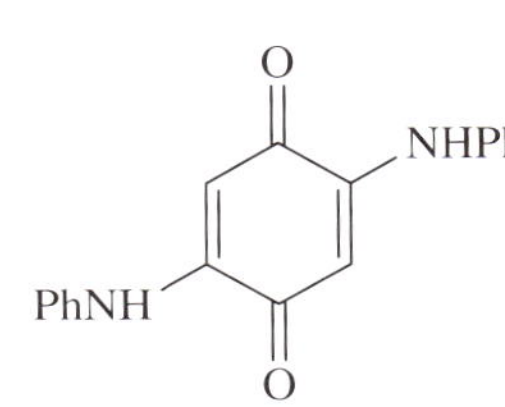

4.7

Aromatic amines react with various carbonyl compounds in a useful series of condensation reactions. Dehydration of the carbinolamines to form imines takes place readily. Reactions of this type are used in heterocyclic synthesis.

4.2.4 Aromatic Diazonium Compounds

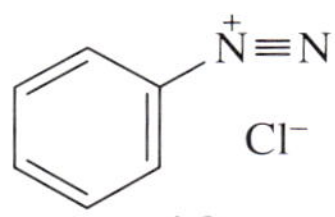

4.8

Aromatic amines react with nitrous acid (sodium nitrite and hydrochloric acid) to form diazonium compounds. In the diazotization reaction the amine behaves as a nucleophile, adding to the nitroso group of the nitrous acid. Loss of water then yields the diazonium compound (**4.8**).

The reactions of diazonium compounds can be divided into two groups: those involving the retention of nitrogen and those involving the loss of nitrogen. The reactions involving the retention of nitrogen are those of an electrophilic nitrogen species. As already mentioned, the diazonium salts will react with an electron-rich aromatic ring such as a phenol by a coupling reaction to give dyestuffs (e.g. **4.9**). Reduction of diazonium compounds with tin(II) chloride gives rise to phenylhydrazines (e.g. **4.10**).

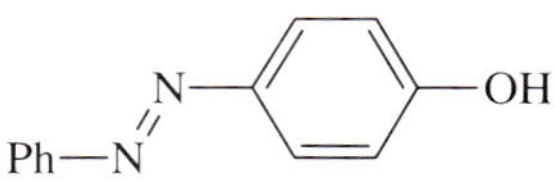

4.9

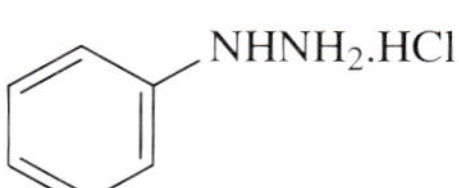

4.10

The reactions involving the displacement of nitrogen, as a molecule of nitrogen gas, are interesting because they lead to a reversal of the reactive character of a specific site on the aromatic ring. Although many of the reactions have a radical character, they involve creating electron deficiency in a σ-bond, allowing attack on the aromatic ring in the σ-bond system by a nucleophilic species. Many of the reactions are catalysed by copper(I) salts and are known as Sandmeyer reactions. There are reactions with halogen, oxygen, nitrogen and carbon nucleophiles (Scheme 4.22). These reactions are of considerable importance in the transformation of aromatic functional groups.

Scheme 4.22

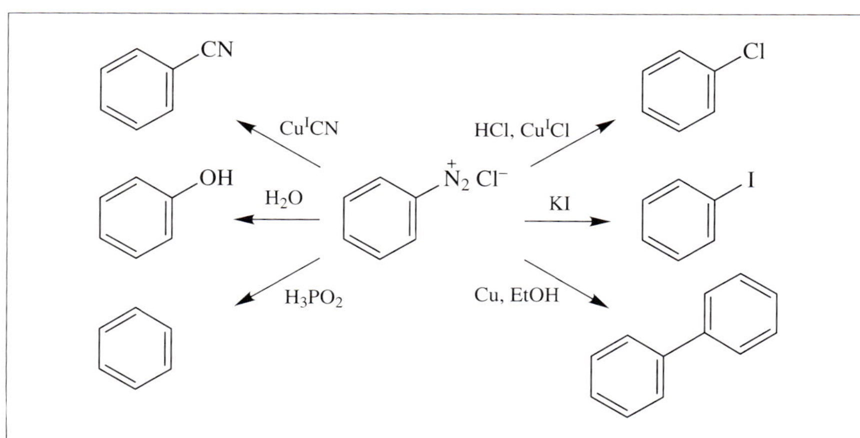

Reduction with phosphinic acid (hypophosphorous acid, H_3PO_2) leads to the replacement of the diazonium group with hydrogen. Thus an amino group can be used in an aromatic sequence to direct a particular aromatic substitution and then it can be removed via the diazonium salt.

4.2.5 Aromatic Nitro Compounds

The aromatic nitro group is a deactivating substituent as far as electrophilic aromatic substitution is concerned. Further electrophilic substitution requires vigorous conditions and takes place at the least deactivated position. Thus the preparation of 2,4,6-trinitrotoluene (TNT) from 4-nitrotoluene requires fuming nitric acid and fuming sulfuric acids. Nitrobenzene is sufficiently unreactive towards electrophilic substitution to be used as a solvent for the Friedel–Crafts alkylation of more reactive aromatic systems.

On the other hand, the aromatic nitro group activates the ring towards nucleophilic addition. The addition may be followed by an elimination. This is exemplified by the preparation of 2,4-dinitrophenylhydrazine from 2,4-dinitrochlorobenzene.

The effect of the nitro group in decreasing the basicity of aromatic amines in the *ortho* and *para* positions, and in increasing the acidity of phenolic hydroxyl groups, has already been mentioned. The nitro group also activates aromatic methyl groups both towards oxidation and towards base-catalysed condensation reactions, as exemplified by the formation of 2,4-dinitrobenzoic acid and the formation of 2,4-dinitrostilbene from 2,4-dinitrotoluene (Scheme 4.23). The carbanion that is formed on the methyl group is stabilized by delocalization onto the nitro group.

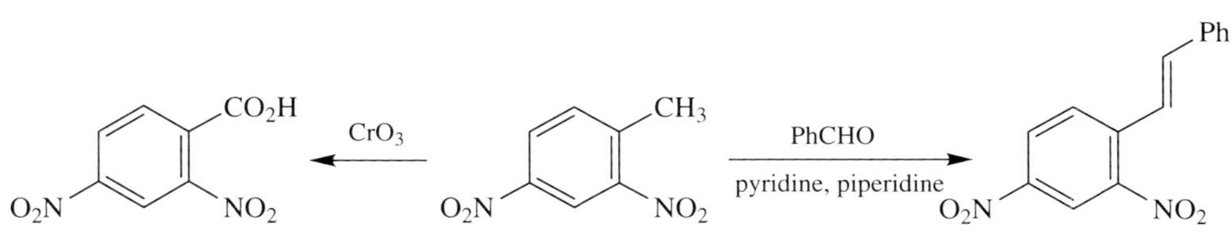

Scheme 4.23

4.2.6 Aromatic Carboxylic Acids

Aromatic carboxylic acids can be prepared by oxidation of the corresponding benzyl alcohols and benzaldehydes. An aromatic methyl group is also sufficiently activated by the aromatic ring for it to be oxidized to the acid under vigorous conditions, such as heating with chromium(VI) oxide. Other conventional methods include the hydrolysis of aromatic nitriles and the carboxylation of aromatic Grignard reagents with carbon dioxide.

Aromatic carboxylic acids show many properties that parallel those of aliphatic carboxylic acids. Thus acyl chlorides are formed on treatment with thionyl chloride and esters are obtained on treatment with the alcohol and an acid catalyst. Benzoate and 3,5-dinitrobenzoate esters are used in the characterization of alcohols and are prepared using the reaction of the acyl halide with the alcohol.

There are a number of reactions which reveal the interaction between the aromatic ring and the carboxyl group. The carboxyl group and the corresponding esters are deactivating groups, and hence nitration and bromination require relatively vigorous conditions and the substituents enter the *meta* position. The strength of the benzoic acids reflect the electron-withdrawing or -donating character of additional substituents on the aromatic ring. Thus an electron-withdrawing *p*-nitro group increases the strength of the acid, whereas an electron-donating *p*-methoxy group weakens the acid (pK_a for benzoic acid, 4.17; for 4-nitrobenzoic acid, 3.43; for 4-methoxybenzoic acid, 4.47). Other properties, such as the rate of hydrolysis of the methyl esters, can be similarly correlated with the nature of the substituents.

Exercise

Copy this revision chart showing the relationships between aromatic compounds, and fill in the relevant reagents and conditions for the inter-relationships beside the arrows. Note the activating, deactivating and orienting effects of the various substituents.

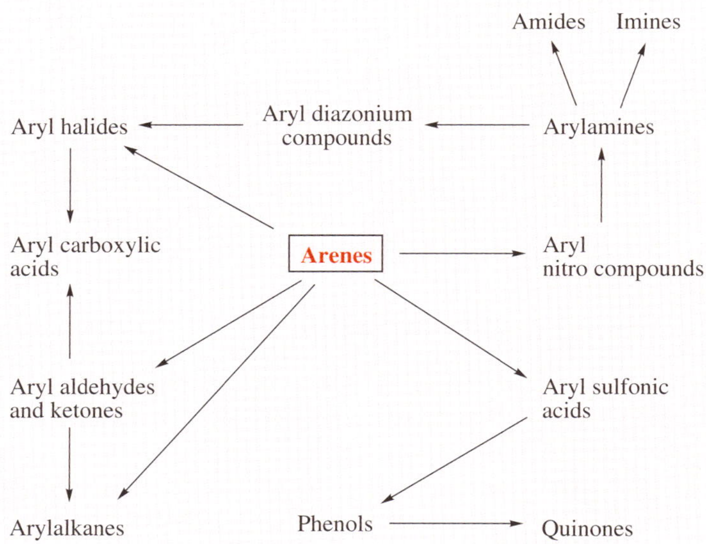

Exercise

Copy this revision chart showing the relationship of aromatic diazonium compounds to other aromatic functional groups, and fill in the relevant reagents and conditions for the inter-relationships beside the arrows.

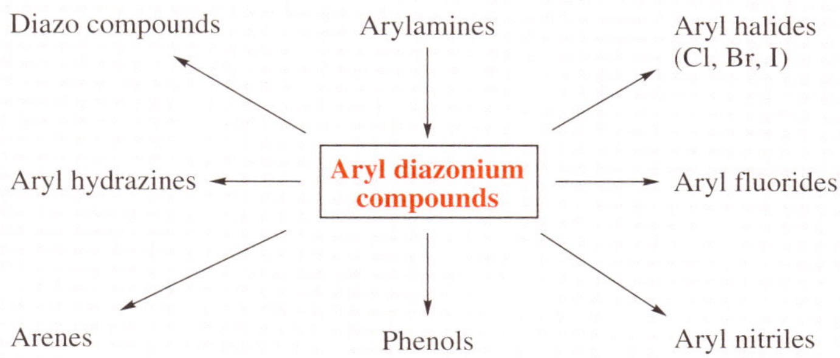

4.3 Heteroaromatic Compounds

Heterocyclic compounds are cyclic structures which contain an atom other than carbon as part of a ring system. **Heteroaromatic compounds** are a group of these which possess a cyclic conjugated system that can satisfy the $(4n + 2)\pi$ electron rule and thus have some aromatic stabilization.

The insertion of a heteroatom such as nitrogen into a benzenoid aromatic ring can be considered to take place in two ways. In **pyridine** (C_5H_5N; see Box 4.2) a nitrogen atom replaces a C–H. The aromatic sextet is then made up of five electrons from the five carbon atoms and one from the sp^2 hybridized nitrogen atom. The lone pair on the nitrogen remains localized on the nitrogen atom and lies in the plane of the ring and it does not participate in the sextet. On the other hand, in the five-membered ring of **pyrrole** (C_4H_5N) a nitrogen atom replaces two carbon atoms, and the aromatic sextet is made up of four electrons from the remaining four carbon atoms and two electrons from the lone pair on the nitrogen.

The heteroatom may be oxygen or sulfur as in the **pyrylium salts**, **furan** or **thiophene**. The extent to which these compounds possess an enhanced stability arising from their aromatic resonance energy varies. There are more complex ring systems containing several heteroatoms such as the **pyrimidine**, **imidazole** and **thiazole** ring systems. Heteroaromatic rings may be fused to a second aromatic ring as in **quinoline**, **isoquinoline** and **indole**.

Box 4.2 Some Heteroaromatic Compounds

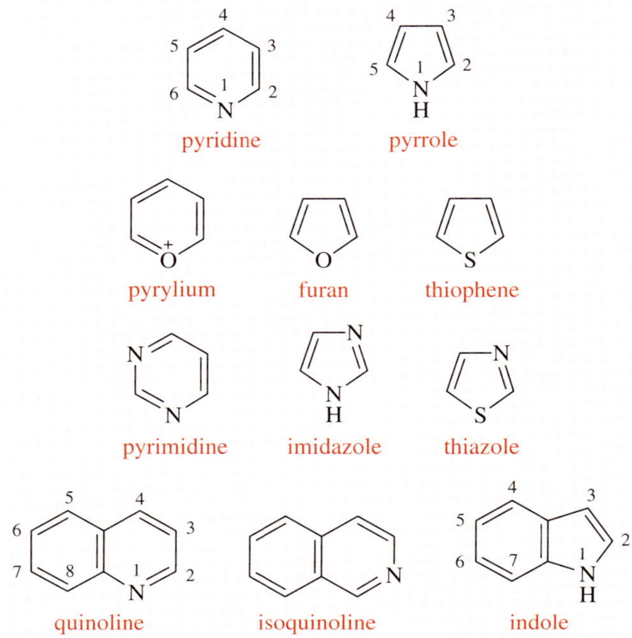

When the heteroaromatic ring is derived by replacing a –CH=CH– by a –CH=N– as in pyridine, the more electronegative nitrogen has an overall electron-attracting effect, as in an imine. It distorts the π-electron cloud of the aromatic ring system, so making the carbon atoms **π-deficient**. Hence, although the nitrogen atom itself reacts readily with electrophiles, the remainder of the ring system is resistant to electrophilic attack. On the other hand, the nitrogen of pyrrole contributes its lone pair to the aromatic sextet. The nitrogen is no longer basic, but the π-system is **electron rich** and there is a parallel to the reactivity of enamines. This division of heteroaromatic compounds into π-deficient and π-excessive compounds provides a useful basis for understanding their chemistry.

4.3.1 Reactions of π-Deficient Heteroaromatic Compounds

The pyridine ring system is the archetypical π-deficient heteroaromatic ring with a resonance energy of 125.5 kJ mol^{-1}. The ring is deactivated towards towards electrophilic attack, but may be attacked by nucleophiles.

Pyridine is only brominated, nitrated or sulfonated under vigorous conditions (Scheme 4.24) with reaction occurring at the least deactivated 3-position. Pyridine does not undergo Friedel-Crafts alkylation or acylation. In many cases the electrophile attacks the nitrogen first to form a pyridinium salt, which is even less reactive towards electrophiles.

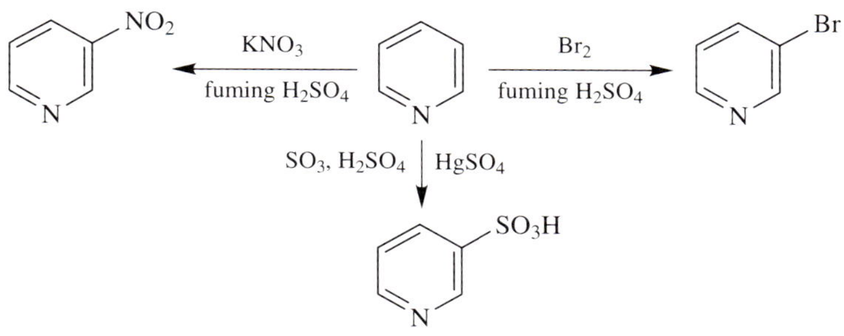

Scheme 4.24

An indication of the deactivating effect of replacing a CH=CH by an imine (CH=N–) is that the electrophilic substitution of quinoline and isoquinoline, in which a benzenoid ring is fused to a pyridine ring, takes place in the benzenoid ring at C-5 and to a lesser extent at C-8 (Scheme 4.25).

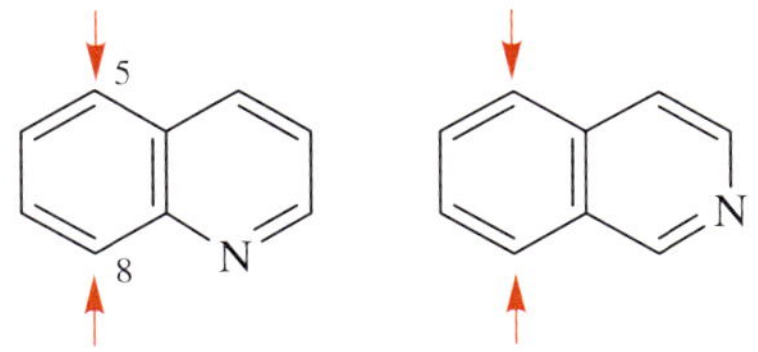

Scheme 4.25

Pyridine reacts with nucleophiles under vigorous conditions at C-2 or C-4 by an addition–elimination mechanism. Thus pyridine reacts with sodamide (the Tschitschibabin reaction; Scheme 4.26a). Pyridine will also react with lithium aluminium hydride to give predominantly a 1,2-dihydropyridine (Scheme 4.26b), whereas sodium in ethanol gives mainly the 1,4-isomer (Scheme 4.26c).

Unlike their benzenoid counterparts, halogen substituents at C-2 and C-4 are sensitive to nucleophilic displacement by an addition–elimination mechanism. There is a parallel to the reaction of imino halides and acyl halides (see Scheme 4.27a). There are a number of other parallels to carbonyl chemistry. Thus a methyl group at C-2 or C-4 undergoes condensation reactions (Scheme 4.27b) and an acetic acid residue at C-2 undergoes decarboxylation (Scheme 4.27c).

Spectroscopic evidence shows that the 2- and 4-hydroxypyridines are

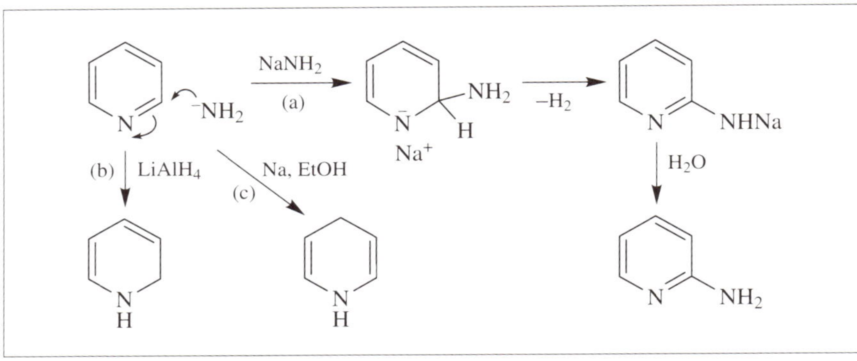

Scheme 4.26

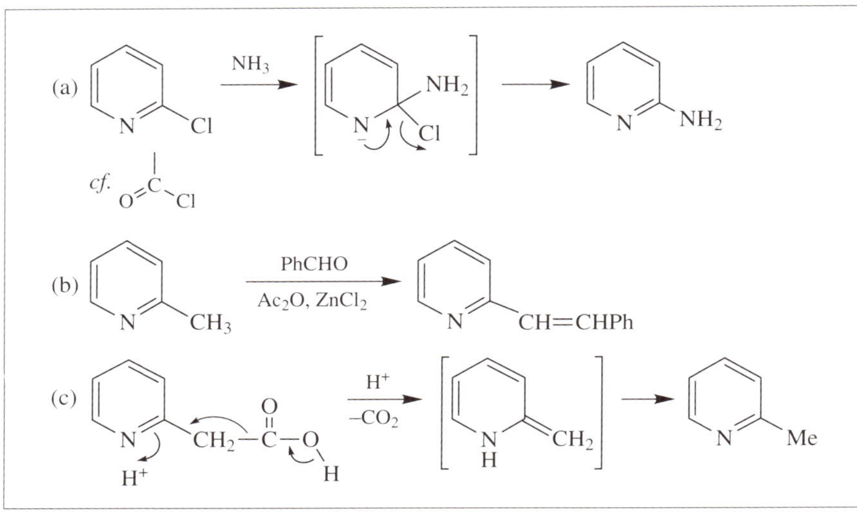

Scheme 4.27

tautomeric with the corresponding **pyridones**, with the latter predominating (Scheme 4.28). These compounds may be alkylated on either the nitrogen or the oxygen, depending on the reaction conditions.

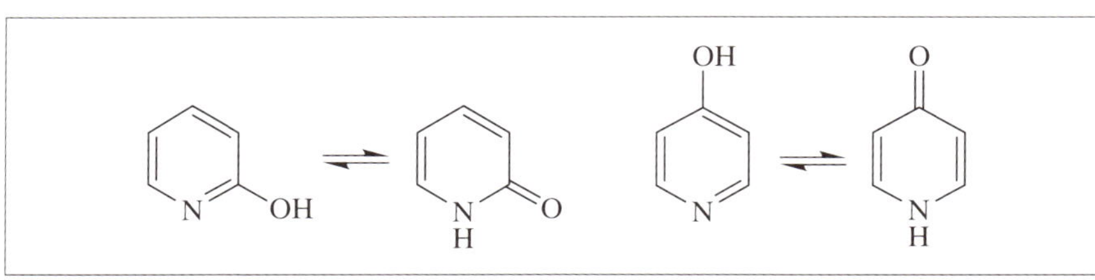

Scheme 4.28

Pyridine is weakly basic. It is less basic than aliphatic amines because the lone pair is in an sp^2 rather than an sp^3 orbital which, with higher 's' character, is held more closely to the nucleus. Nevertheless, alkyl and reactive aryl halides react with pyridines to form quaternary *N*-alkyl (or *N*-aryl) salts.

Pyridine forms a number of complexes and salts which are useful synthetic reagents. These include pyridinium hydrobromide perbromide ($C_5H_5NH^+Br_3^-$), which is a convenient crystalline, easily handled, brominating agent, the pyridine–sulfur trioxide complex ($C_5H_5N{:}SO_3$) for sulfonations, pyridine–borane ($C_5H_5N{:}BH_3$) for hydroborations and the pyridine–chromium(VI) oxide complex ($C_5H_5N{:}CrO_3$) for oxidations.

The oxidation of pyridinium salts with alkaline potassium ferricyanide provides a facile route to *N*-substituted pyridones (Scheme 4.29a), although the sensitivity of pyridinium salts to nucleophilic attack may lead to ring fission reactions (Scheme 4.29b).

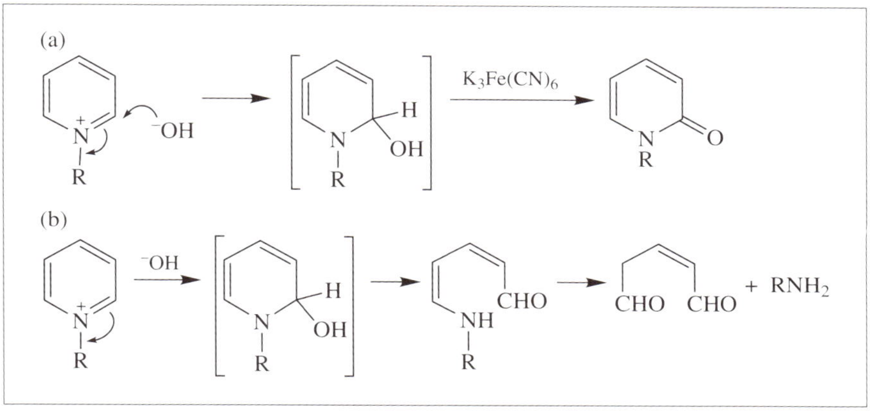

Scheme 4.29

Oxidation of pyridine with a peroxy acid gives **pyridine *N*-oxide**. The *N*-oxides may be reduced to regenerate the parent pyridine with a variety of reagents, including phosphorus(III) compounds. Their value as synthetic intermediates is that there is a back-donation from the oxygen to activate the 2- and 4-positions towards electrophilic attack. Thus nitration of pyridine *N*-oxide, although still requiring vigorous conditions (conc. HNO_3/conc. H_2SO_4), gives 4-nitropyridine *N*-oxide (**4.11**).

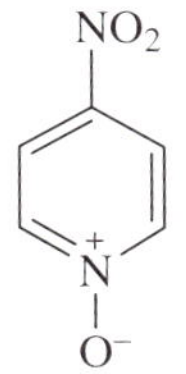

4.11

4.3.2 Reactions of π-Excessive Heteroaromatic Compounds

Pyrrole is a typical π-excessive heteroaromatic compound, with a resonance energy of 103 kJ mol^{-1}. While thiophene possesses rather more aromatic character, there is less stabilization associated with furan.

The involvement of the nitrogen lone pair in the aromatic sextet means that the nitrogen of pyrrole is non-basic. When the pyrrole ring system is protonated, addition takes place at C-2 (Scheme 4.30a). Electrophilic attack on pyrrole is rapid and also takes place at C-2. Reaction at C-3 is less favoured (Scheme 4.30b).

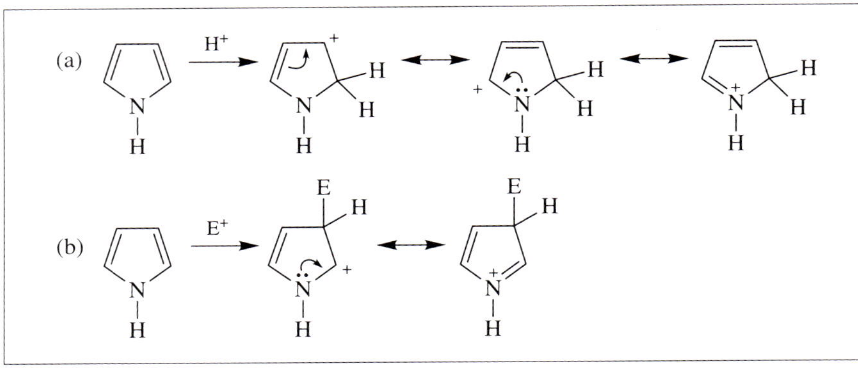

Scheme 4.30

Pyrrole and its simple alkyl derivatives are polymerized by many of the typical acidic electrophiles. Nitration with acetyl nitrate (Scheme 4.31) under mild conditions gives the 2-nitro compound, together with small amounts of 3-nitropyrrole. Halogenation is rapid but it gives a complex mixture of polyhalogeno compounds.

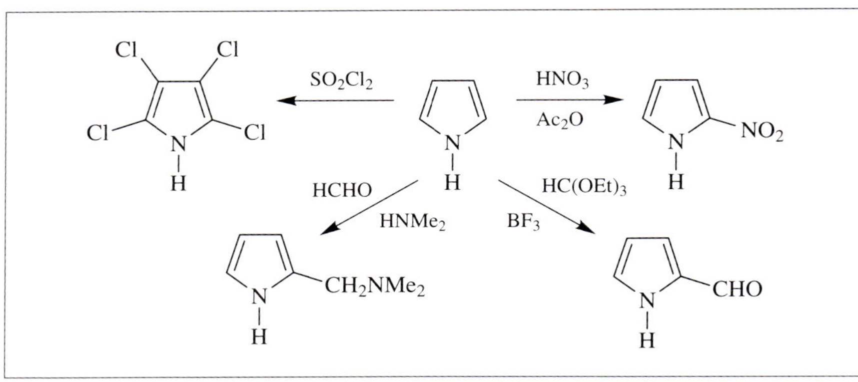

Scheme 4.31

Pyrrole can be easily acylated and alkylated, reacting as an enamine. Formylation takes place at the C-2 position with ethyl orthoformate and boron trifluoride etherate as a catalyst, with hydrogen cyanide and hydrogen chloride (Gatterman reaction) or with dimethylformamide and phosphorus oxychloride (Vilsmeier reaction). The products of these reactions are useful for coupling pyrroles to give dipyrrolylmethanes and thence the tetrapyrroles of the porphyrin and corrin ring systems. The acid-catalysed condensation of pyrrole with 4-dimethylaminobenzaldehyde (the Ehrlich test for pyrroles) is a colour test for the ring system which follows this pattern (Scheme 4.32).

Although the nitrogen is non-basic, the N–H hydrogen is acidic and, in the presence of a strong base, pyrrole forms a powerful nucleophile. Pyrrole reacts with Grignard reagents to give a pyrrolylmagnesium halide (Scheme 4.33a), releasing a hydrocarbon. The pyrrole anion is resonance

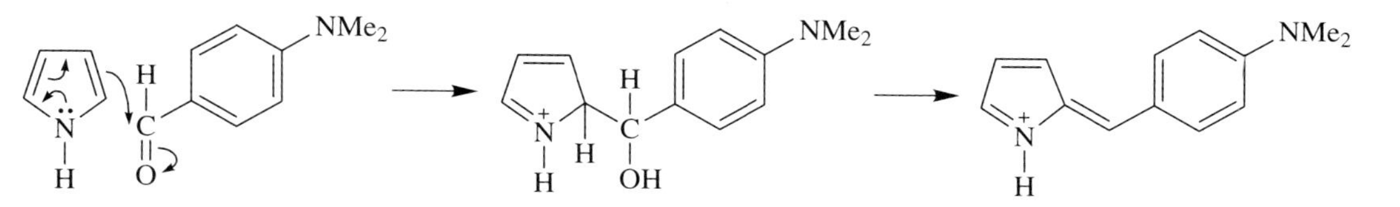

Scheme 4.32

stabilized (Scheme 4.33b) and will react either on nitrogen or carbon (Scheme 4.33c).

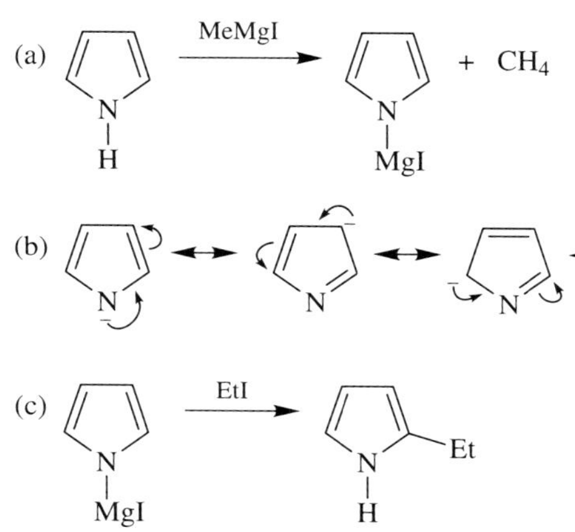

Scheme 4.33

Thiophene is closer to benzene in its resonance energy (116 kJ mol^{-1}) and aromatic character. It is somewhat more stable to acidic conditions and undergoes, for example, Friedel–Crafts acylation.

On the other hand, furan has less resonance stabilization (91 kJ mol^{-1}) and undergoes a number of reactions in which the aromatic character is lost. For example, typical of a diene, it undergoes a Diels–Alder reaction with maleic anhydride (Scheme 4.34).

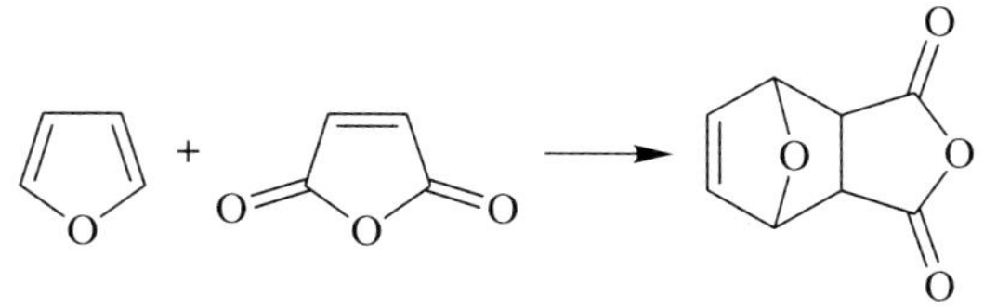

Scheme 4.34

Summary of Key Points

1. Aromaticity is associated with the presence of a cyclic conjugated system containing $(4n + 2)\pi$ electrons.

2. The reaction of an aromatic ring with an electrophile involves an addition to form a Wheland intermediate. This is followed by the elimination of a proton and reversion to the aromatic system.

3. Substituents which are electron donating, and can stabilize a positive charge, activate the ring towards electrophilic attack and direct further substitution to the *ortho* and *para* positions. Substituents which are electron withdrawing deactivate the ring towards electrophilic attack and direct further substitution to the *meta* position.

4. Electrophiles for aromatic substitution include the halonium ion, the nitronium ion and the carbonium ion. The latter may be generated from alkyl and acyl halides using Lewis acid catalysis in the Friedel–Crafts reactions.

5. Nitro substituents activate halogens in the *ortho* and *para* positions to nucleophilic substitution by an addition–elimination mechanism.
6. The aromatic ring of a phenol is activated towards electrophilic attack, while the phenolic hydroxy group is more acidic than an aliphatic alcohol.
7. Aromatic amines are less basic than aliphatic amines and the basicity is influenced by substituents on the aromatic ring. The reactivity of amines is modified by conversion to amides such as acetanilide. These still undergo electrophilic attack in the *ortho* and *para* positions.

8. Treatment of an aromatic amine with nitrous acid gives diazonium compounds. These couple with phenols to give dyes. Displacement of nitrogen from a diazonium compound by a halide or cyanide ion takes place with copper catalysis and is a useful synthetic method.

9. Heteroaromatic compounds may be divided into π-deficient (*e.g.* pyridine) and π-excessive (*e.g.* pyrrole) classes.

10. The pyridine ring is deactivated towards electrophilic attack but may be attacked by nucleophiles. Electrophilic attack occurs on the nitrogen and on the ring at the least deactivated 3-position. Halogen substituents at C-2 and C-4 are sensitive to nucleophilic displacement by an addition–elimination mechanism.

11. Pyrrole is non-basic. The pyrrole ring is very sensitive to acid. Electrophilic attack on pyrrole is rapid and takes place predominantly at C-2. The N–H hydrogen of pyrrole is acidic and, in the presence of a strong base, pyrrole forms a powerful nucleophile.

Worked Problems

Q Explain why diazocyclopentadiene (**1**) is more stable than typical diazoalkanes.

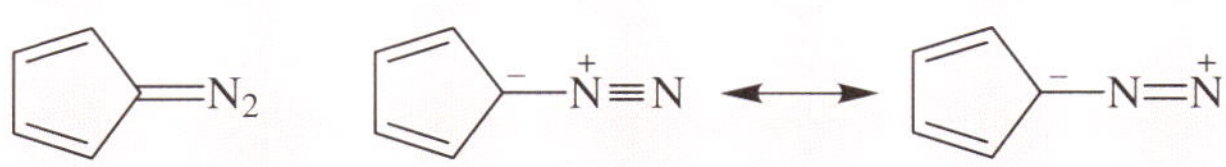

1 **2** **3**

A Diazocyclopentadiene can be written in the dipolar forms **2** and **3**. The cyclopentadienyl anion possesses six π-electrons and hence it has some aromatic stabilization.

Q Devise a synthesis of *o*-bromophenol.

A Under normal conditions, phenol brominates rapidly to give 2,4,6-tribromophenol and even under mild conditions with bromine in acetic acid it gives *p*-bromophenol. Consequently, the reactivity of the ring has to be diminshed and the *para* position protected. In order to achieve this, phenol is first sulfonated to give the 2,4-disulfonate. This is then monobrominated and the sulfonate groups are removed by heating with water.

Q Treatment of nitrobenzene with potassium *t*-butoxide and *t*-butyl hydroperoxide gives *p*-nitrophenol. Provide a mechanism for this reaction.

A A nitro group activates the *ortho* and *para* positions towards nucleophilic addition. The adduct from the *t*-butyl hydroperoxide anion may then fragment with the loss of *t*-butyl alcohol:

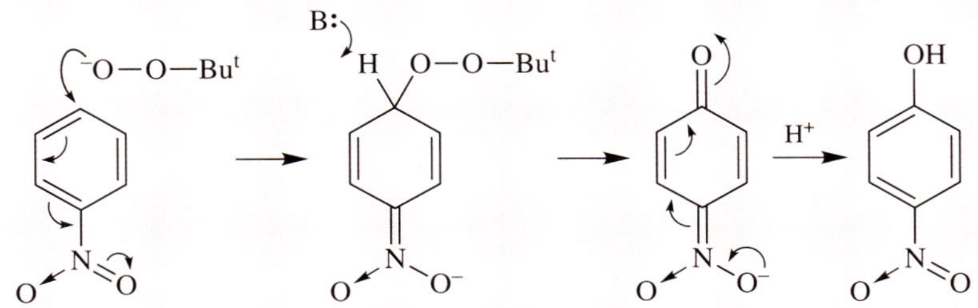

Problems

4.1. Which of the following compounds could be expected to show aromatic properties?

(a)  (b)  (c) 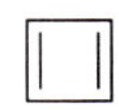 (d) 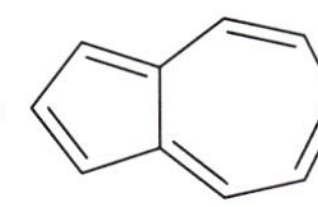

4.2. Predict the sites of mononitration of the following compounds:

(a) 

(b) 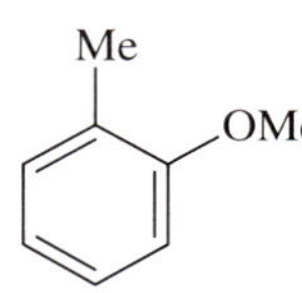

(c) 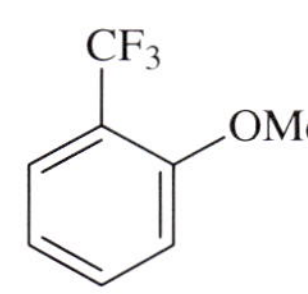

(d) 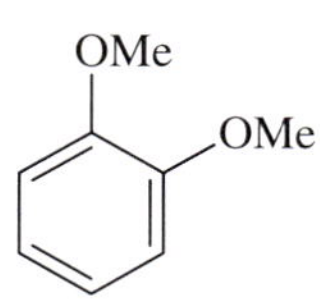

(e) 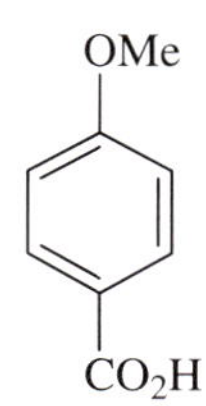

(f) 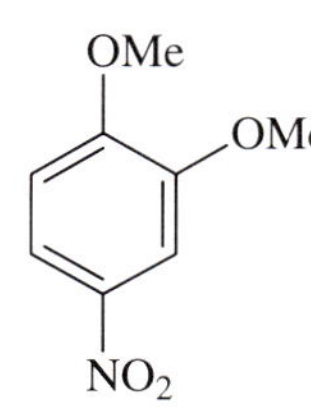

(g) 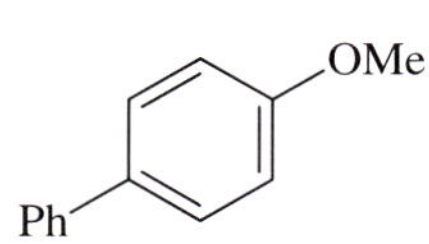

(h) 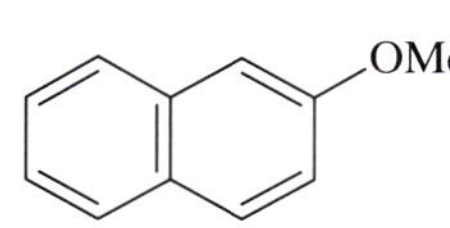

(i) 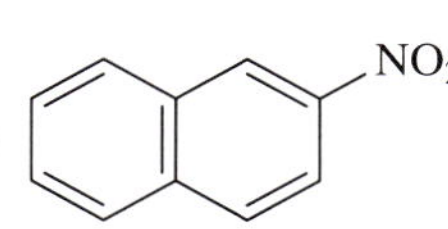

4.3. Show how the following compounds might be prepared from a suitable monosubstituted aromatic compound *via* the transformation of a diazonium salt.

(a) 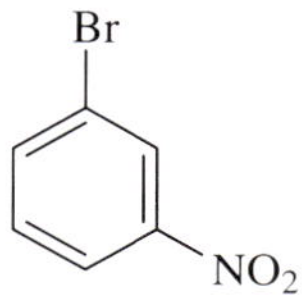 (b) 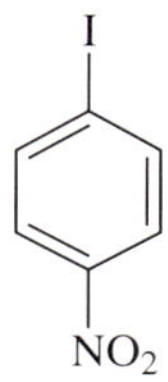 (c) 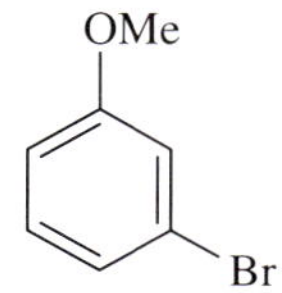 (d) 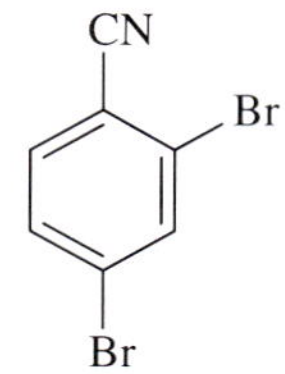

4.4. How might the deuterium label be introduced into the following compounds:

(a) 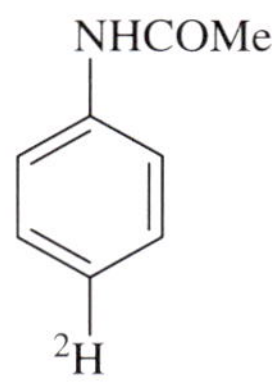 (b) 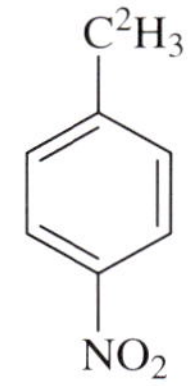

4.5. Suggest routes for the preparation of the following compounds. Give your reason for the selection of the starting material.

(a) 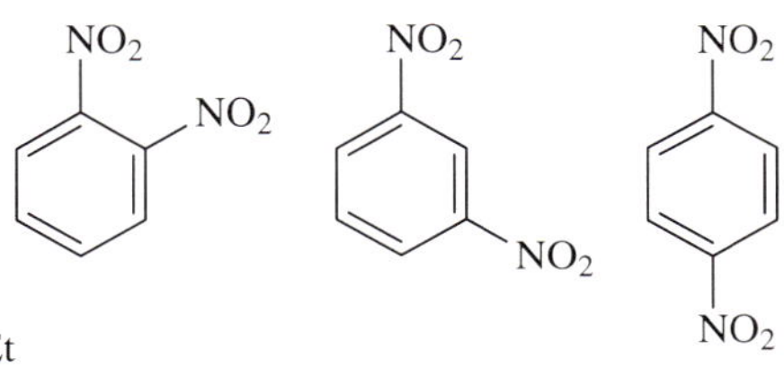 (from benzene)

(b) 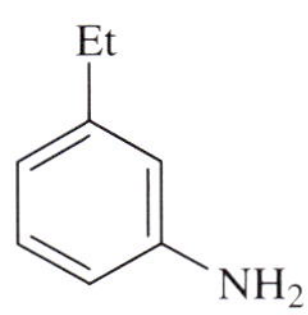 (from benzene) (c) 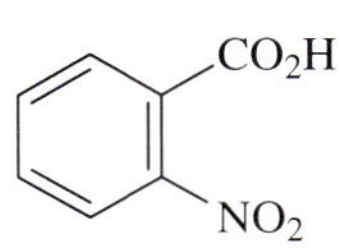 (from toluene)

(d) 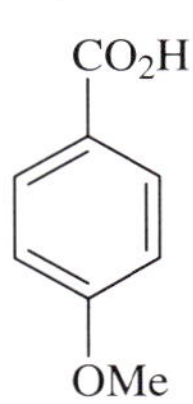 (from either 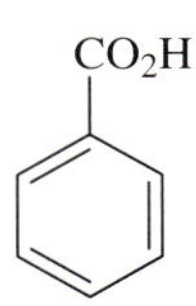 or 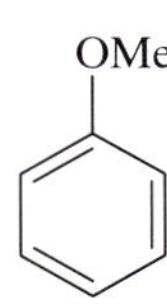)

4.6. Starting with a monosubstituted benzene, show how the following compounds might be prepared. Indicate the reason for your choice of starting material and the sequence of operations.

(a) 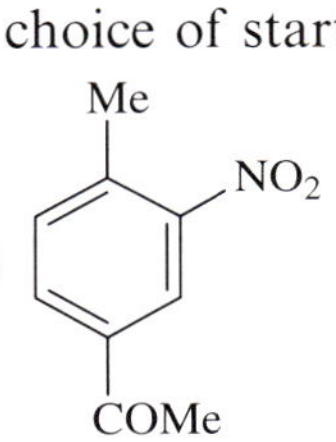 (b) 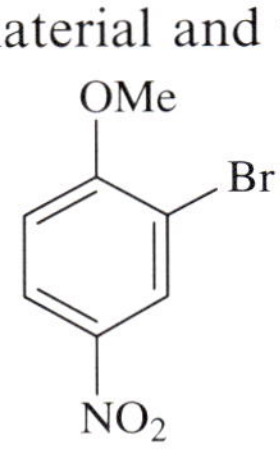 (c) 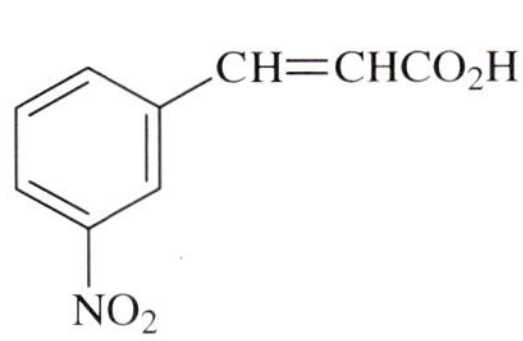

4.7. Write down the formulae of the products of the reaction of *p*-cresol (**4**) with each of the following reagents: (a) toluene-4-sulfonyl chloride and dil. sodium hydroxide; (b) bromine water; (c) benzenediazonium chloride; (d) chloroform and aq. sodium hydroxide at 70 °C; (e) carbon dioxide and aq. sodium hydroxide at 125 °C under pressure; (f) alkaline potassium ferricyanide.

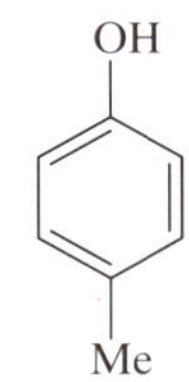

4

4.8. (a) Which of the following is the strongest acid:

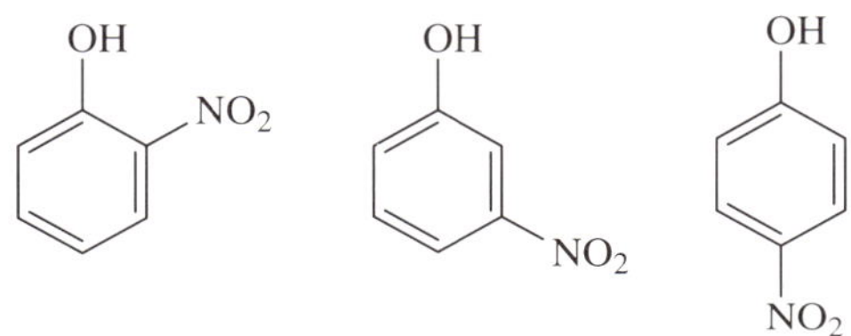

(b) Why is **5** less basic than **6**?

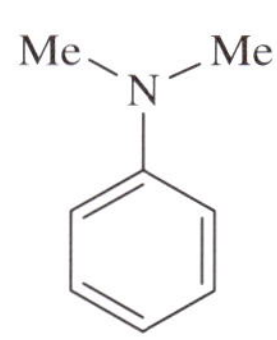

5

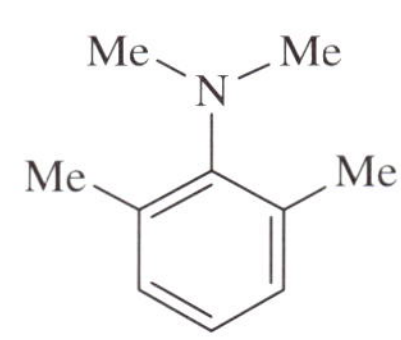

6

(c) Why are picric acid (**7**) and barbituric acid (**8**) acidic ?

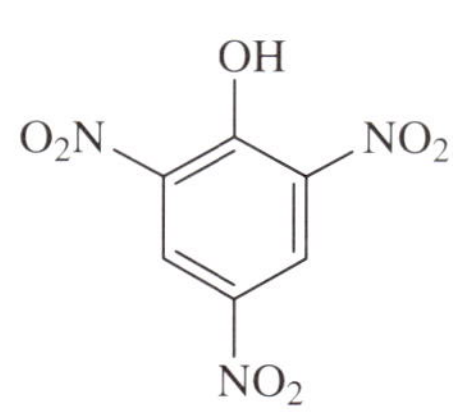

7

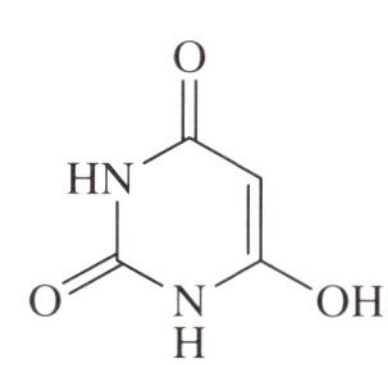

8

4.9. Explain the following observations. (a) Nitration of acetanilide gives mainly 4-nitroacetanilide and nitration of 2-

methylacetanilide gives 2-methyl-4-nitroacetanilide, whilst nitration of 2,6-dimethylacetanilide gives 2,6-dimethyl-3-nitroacetanilide. (b) Treatment of 4-nitroveratrole (**9**) with refluxing aqueous sodium hydroxide affords 2-methoxy-4-nitrophenol, whilst heating with dilute sulfuric acid gives 2-methoxy-5-nitrophenol.

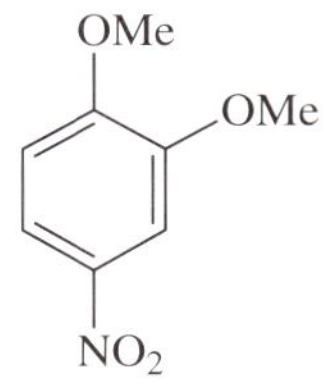

9

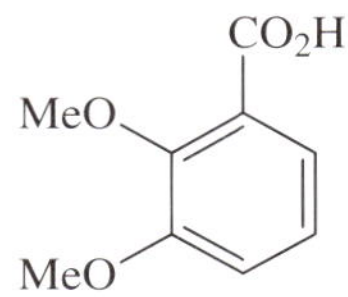

10

4.10. Treatment of *o*-veratric acid (**10**) with formaldehyde and concentrated hydrochloric acid gives the neutral compound meconine, $C_{10}H_{10}O_4$. Meconine is soluble in refluxing alkali and is regenerated from this solution on treatment with hydrochloric acid. What is the structure of meconine?

4.11. Suggest reagents and reactants for the steps in the following syntheses:

(a)

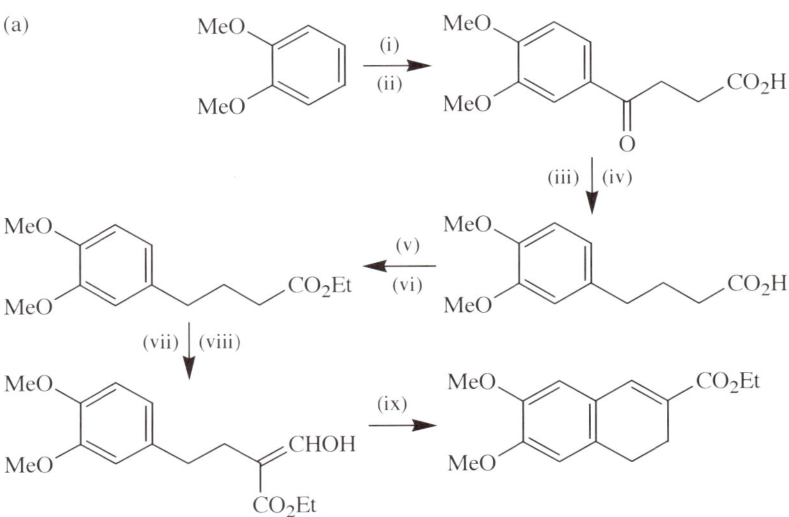

(b)

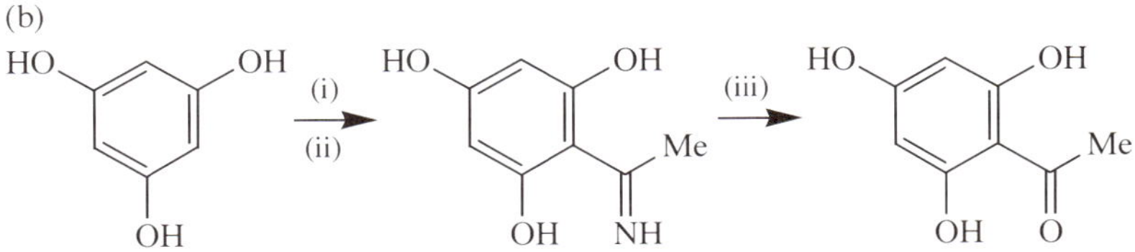

(c)

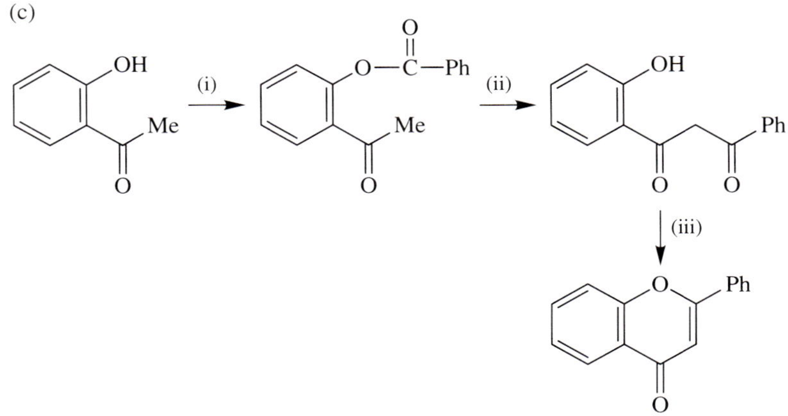

4.12. When phenylhydrazine (**11**) reacts with ethyl acetoacetate, why is compound **12** formed rather than **13**?

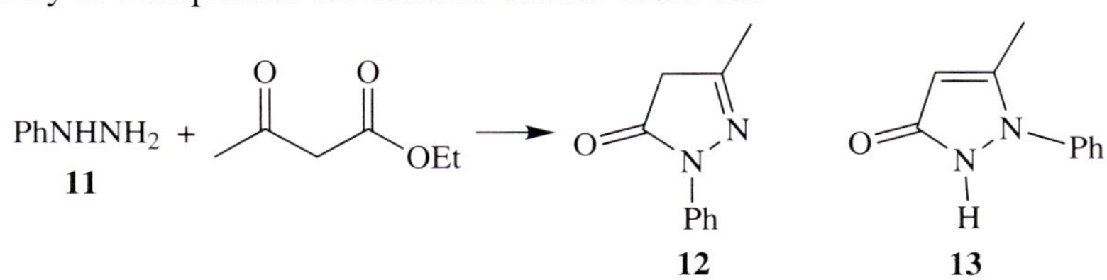

4.13. Identify the products of the following reactions:

(a)

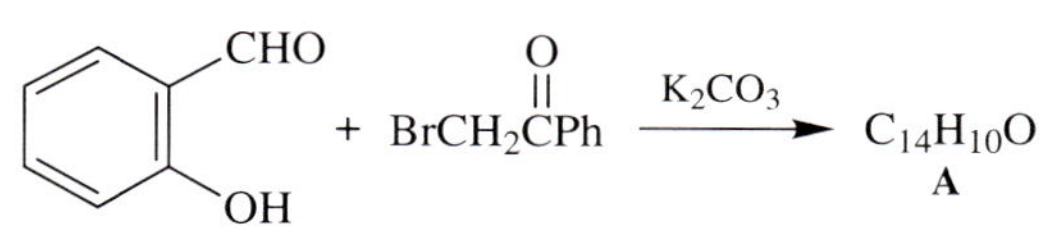

(b)

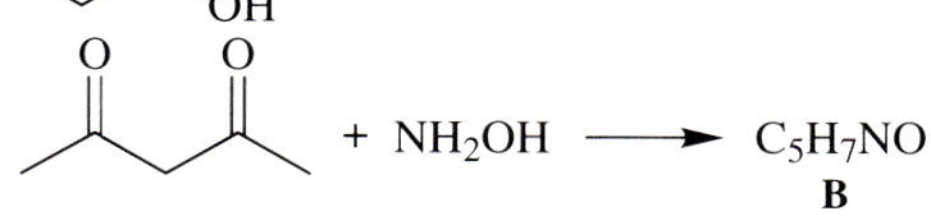

(c)

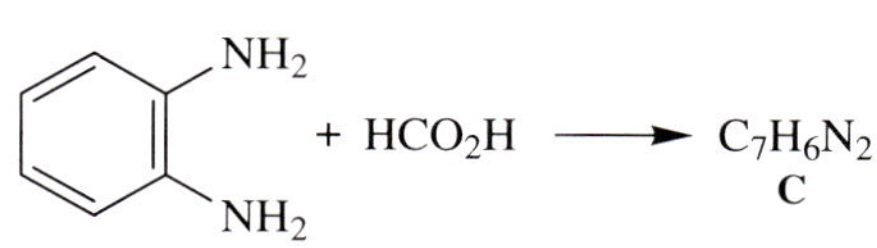

(d)

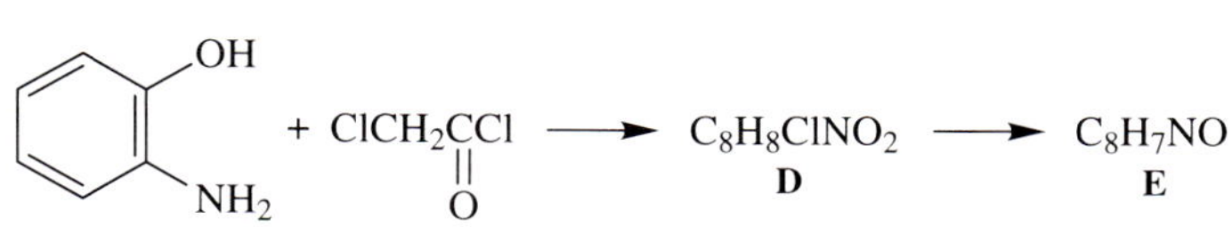

Answers to Problems

Chapter 1

1.1.

(i) (a) 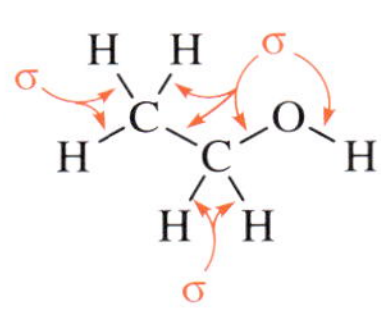 (b) 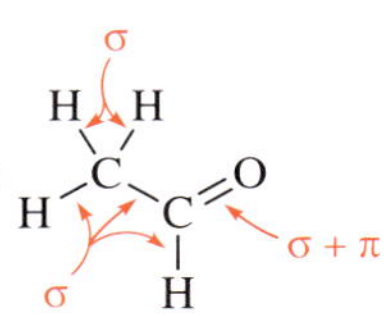 (c) 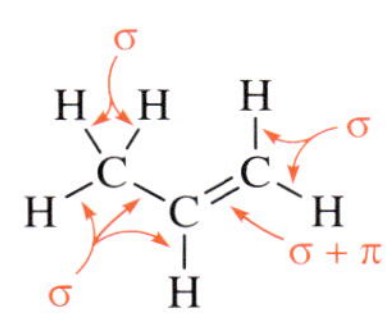

(ii) (d) 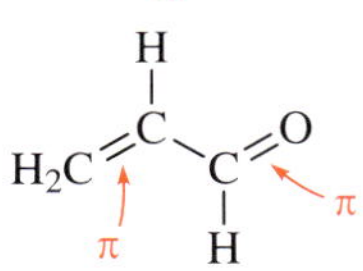 (e) 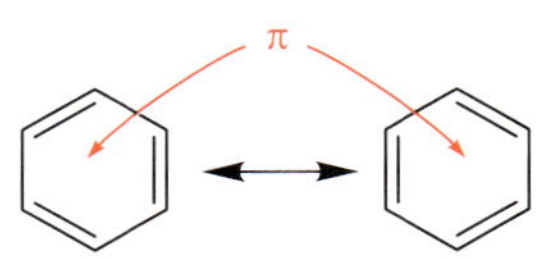

The π-systems in (d) and (e) are conjugated. The π-electron cloud is delocalized over the participating atoms.

1.2.

(a)

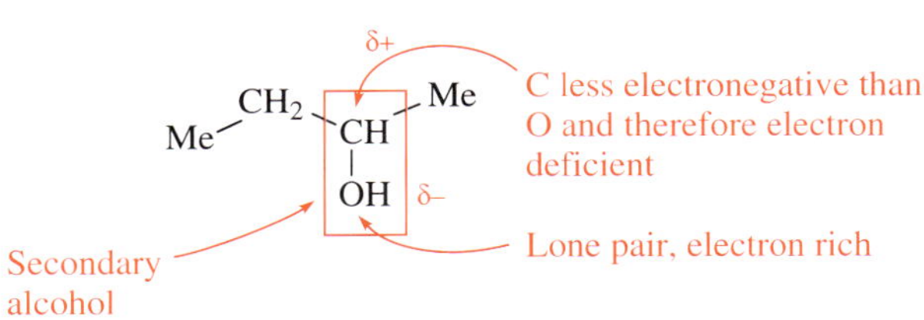

(b)

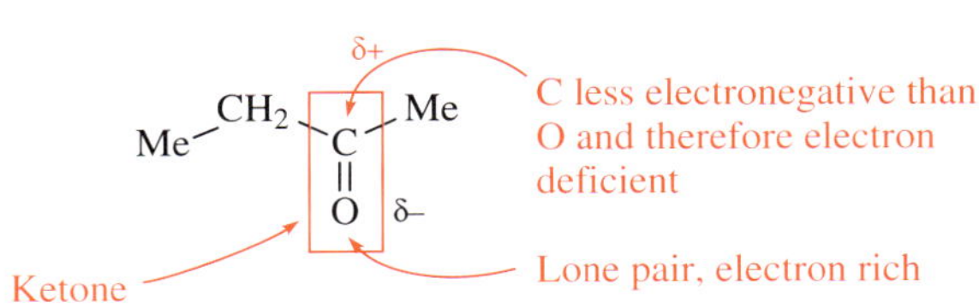

(c)

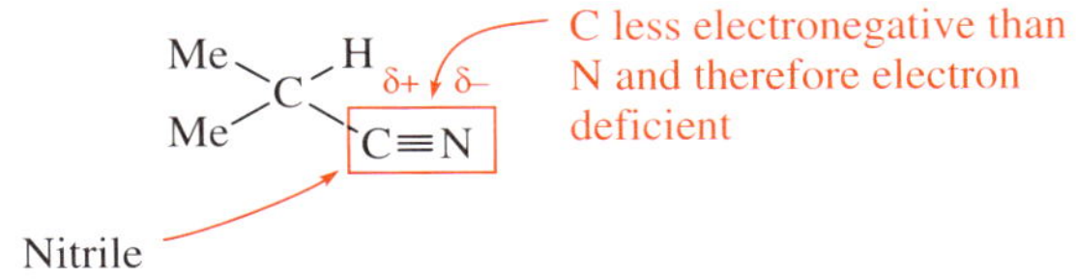

(d)

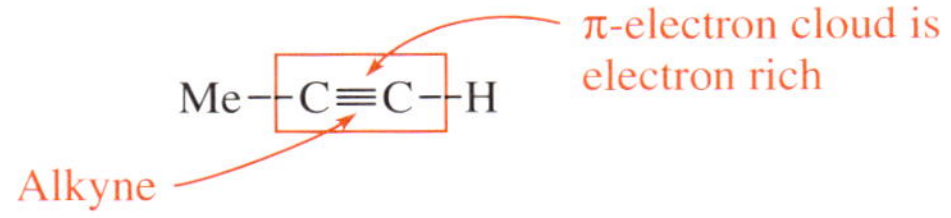

(e)

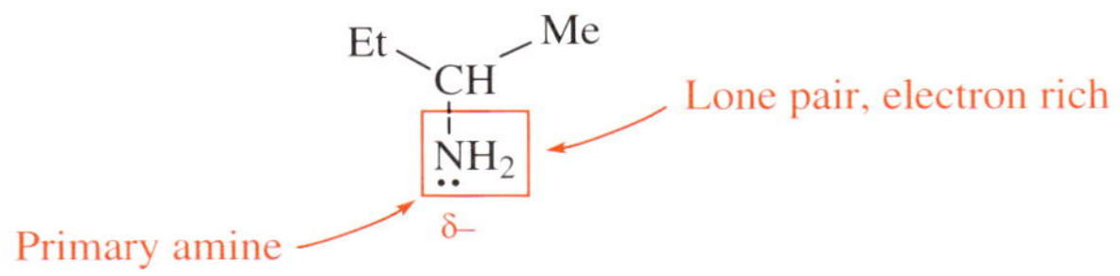

(f)

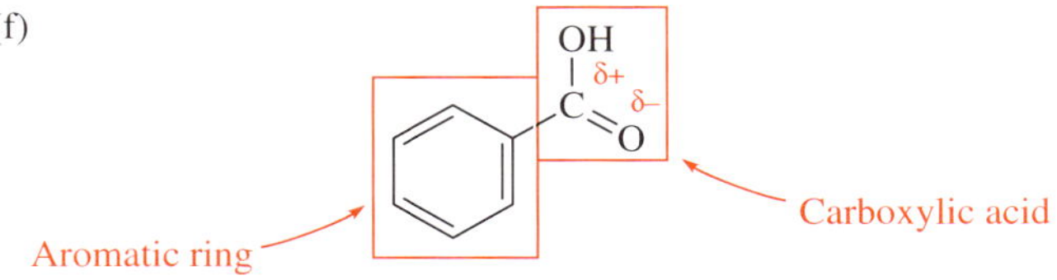

The π-electron system of the aromatic ring is electron rich, but this is subjected to the electron-withdrawing effect of the carboxylic acid.

(g)

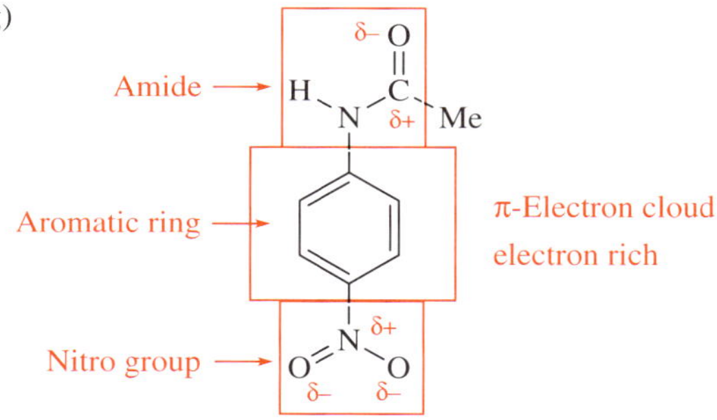

The carbonyl group of the amide withdraws the lone pairs from the nitrogen, making it less basic and less nucleophilic than the parent amine. The two oxygen atoms of the nitro group are more electronegative than the nitrogen and create an electron-withdrawing group. These substituents affect the electron-rich character of the aromatic π-system. The amide is electron donating and the nitro group is electron withdrawing.

(h)

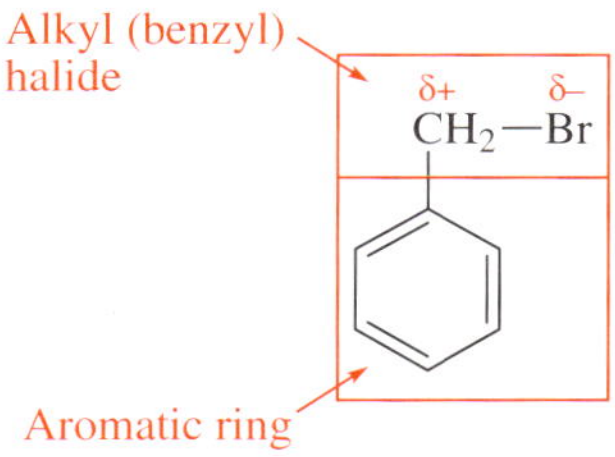

The Br is more electronegative than C and hence the bond is polarized as shown. The π-electron cloud is electron rich, so the aromatic π-system can stabilize an adjacent carbocation and hence the benzylic halide is more reactive than a simple alkyl halide.

(i)

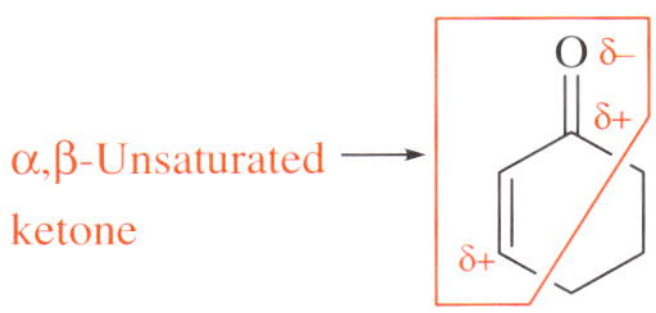

The electron-withdrawing carbonyl group is conjugated to the alkene, making it electron deficient and deactivating it towards electrophilic attack. On the other hand, the π-carbon atom becomes sensitive to nucleophilic attack.

1.3. (a) 3-Methylhexane; (b) 3-methylpent-2-ene; (c) butan-2-one; (d) ethyl ethanoate (ethyl acetate); (e) *N*-methyl-1-propylamine; (f) *N*-methyl-2-propylamine; (g) 3-methylbut-2-enoic acid; (h) 1,2-dimethyl-3-nitrobenzene; (i) 1-bromocyclohex-2-ene.

1.4. (a) Electrophile (bromonium ion); (b) nucleophile (bromide ion); (c) nucleophile (cyanide ion); (d) electrophile (nitronium ion); (e) nucleophile (nitrite ion); (f) nucleophile (amide ion); (g) nucleophile (acetylide ion); (h) electrophile (acylium ion).

1.5.

(a) $\mathrm{Me{-}CH_2{-}\overset{+}{C}H{-}O{-}H \longleftrightarrow Me{-}CH_2{-}CH{=}\overset{+}{O}{-}H}$

(b)

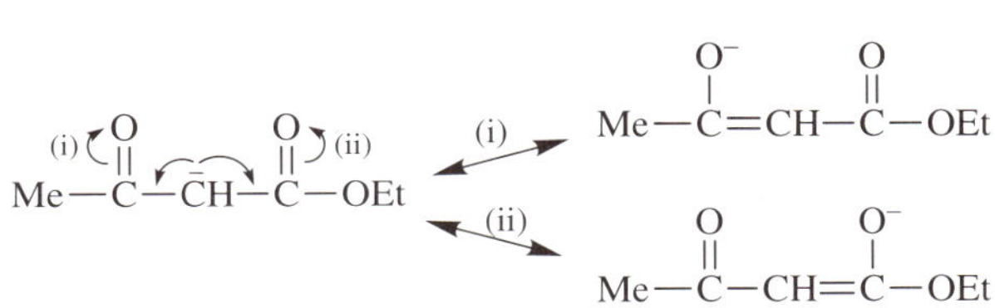

(c)

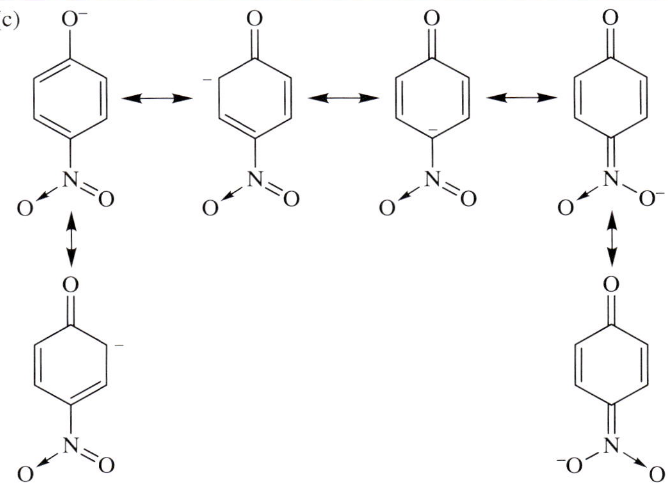

(d)

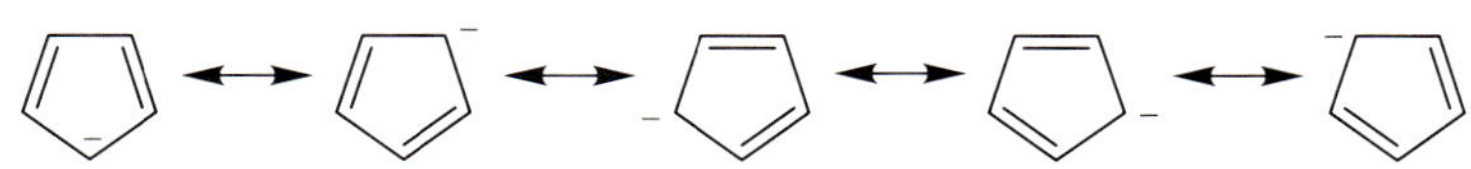

1.6.

(a)

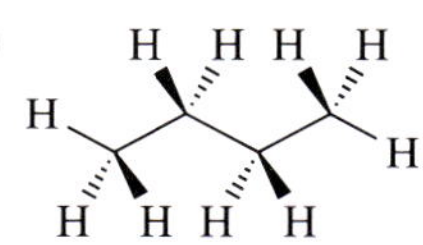

(b)

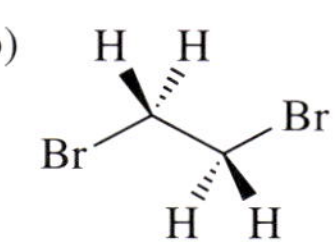

(c)

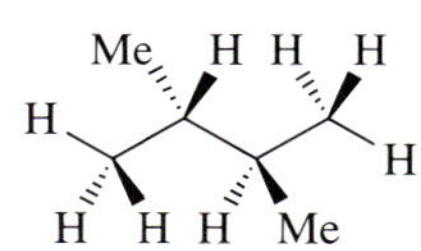

(d)

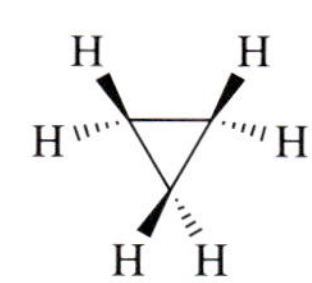

(e)

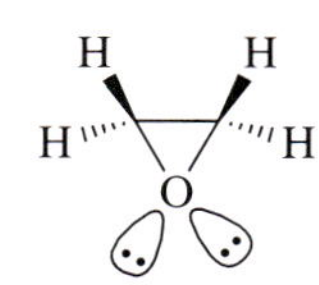

(f)

cis

trans

(g)

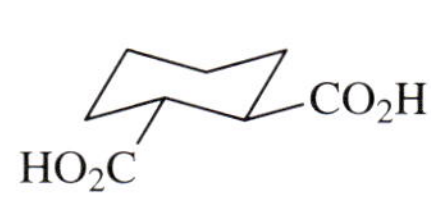

(h)

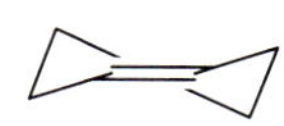

Chapter 2

2.1. (a) Me_2CHCH_2CN; (b) $Me_2CHCH_2NC + Me_2CHCH_2CN$; (c) Me_2CHCH_2OCOMe; (d) $Me_2CHCH_2C{\equiv}CH$; (e) Me_2CHCH_2OH; (f) $Me_2C{=}CH_2$; (g) A = $Me_2CHCH_2PPh_3^+I^-$; B = $Me_2CHCH^-PPh_3^+$

(h) C = 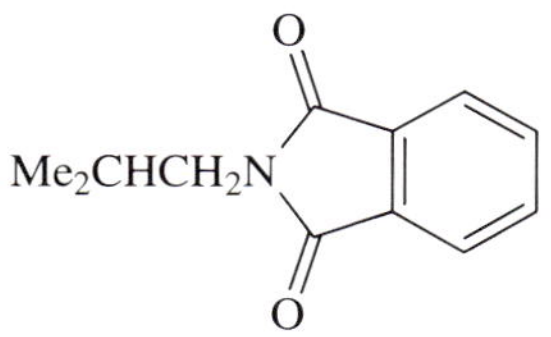 D = $Me_2CHCH_2NH_2$

(i) E = 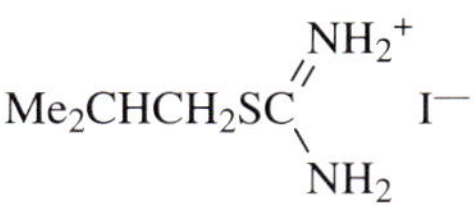 F = Me_2CHCH_2SH

2.2. A = $MeCH_2CH_2Br$; B = $MeCH{=}CH_2$; C = $(MeCH_2CH_2)_2O$; D = Me_2CHBr; E = $MeCO_2H$.

2.3.

(a) $PhCH{=}CH_2 \xrightarrow{^2H_3O^+} PhCH(OH)CH_2{}^2H$

(b) $PhCH{=}CH_2 \xrightarrow[\text{2. } H_2O_2\text{, NaOH}]{\text{1. } B^2H_3\text{.THF}} PhCH^2HCH_2OH$

$B_2{}^2H_6$ can be prepared from NaB^2H_4 and BF_3:Et_2O.

(c) $PhCH{=}CH_2 \xrightarrow{H_3O^+} PhCH(OH)Me \xrightarrow{CrO_3} PhCOMe \xrightarrow{NaB^2H_4} PhC^2H(OH)Me$

(d) $PhCH{=}CH_2 \xrightarrow{HCl} PhCHClMe \xrightarrow{K_2CO_3\text{, } H_2{}^{18}O} PhCH(^{18}OH)Me$

2.4.

(a) A = $MeCH_2CH_2O^-Na^+$; B = $MeCH_2CH_2OEt$

via 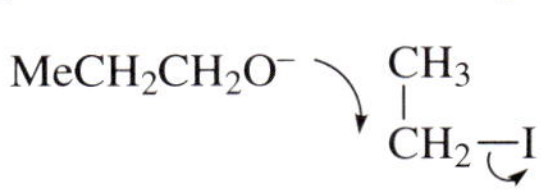

(b) C = $Me_2CHOCOMe$

via 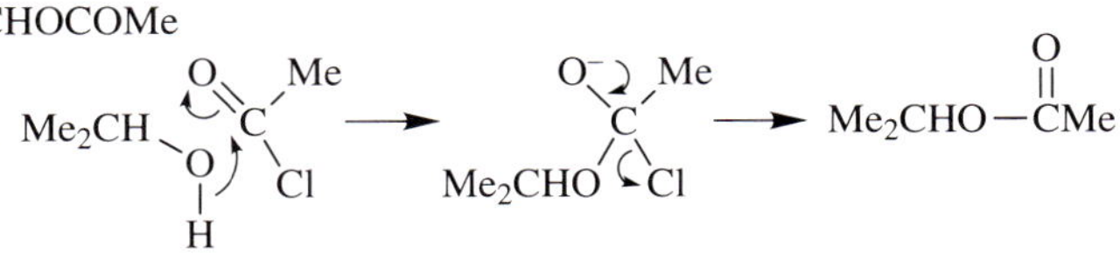

(c) D = Me_2CO

via 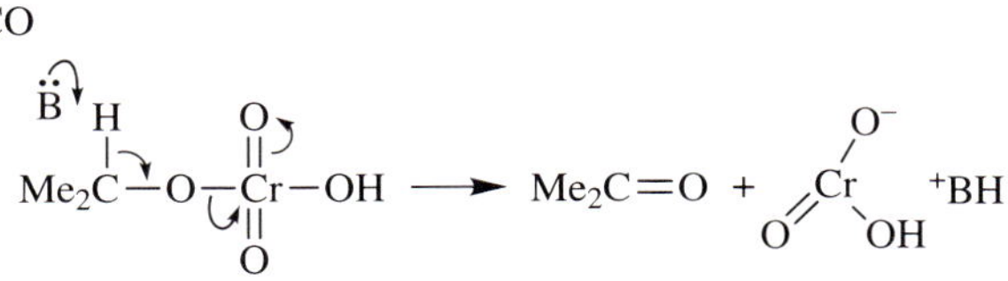

(d) E =

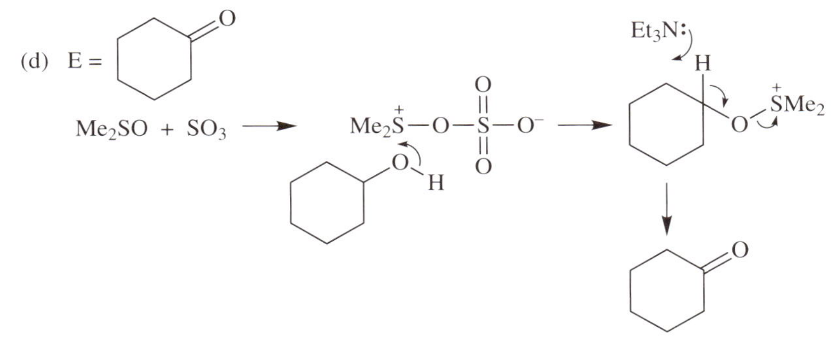

(e)

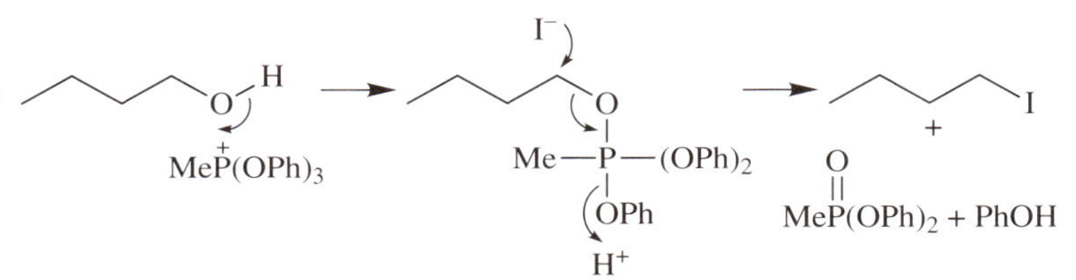

2.5. A = $MeCH_2CH_2CH_2OH$; B = $MeCH_2CH_2CH_2Cl$; C = $MeCH_2CH{=}CH_2$; D = $MeCH_2CO_2H$; E = $MeCH_2CH(Cl)Me$.

2.6. (a) Hydrolysis of the epoxycyclopentene with acid gives a resonance-stabilized allyl carbocation which may react with a hydroxyl group at either the 1- or 3-position:

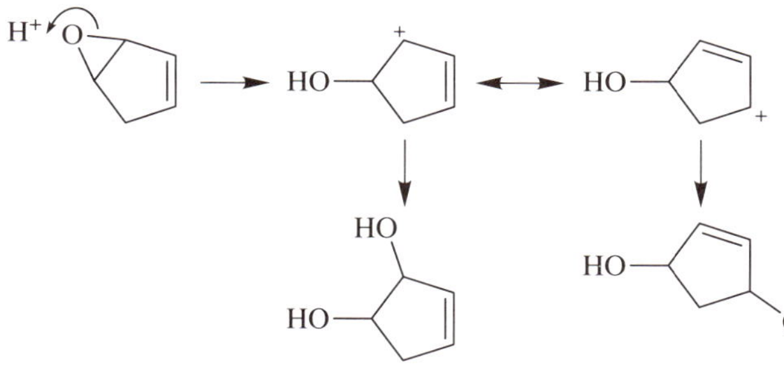

(b) A compound which forms a 2,4-dinitrophenylhydrazone and is oxidized to an acid with the same number of carbon atoms is an aldehyde. Cleavage of the epoxide gives a benzylic carbocation with sufficient lifetime to permit a rearrangement to take place:

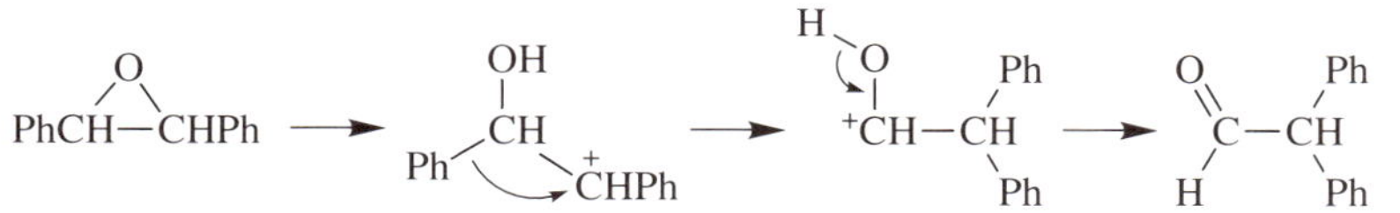

2.7.

(a)

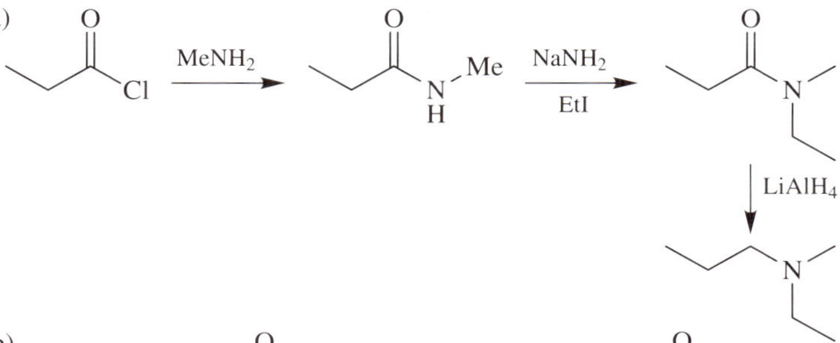

(b)

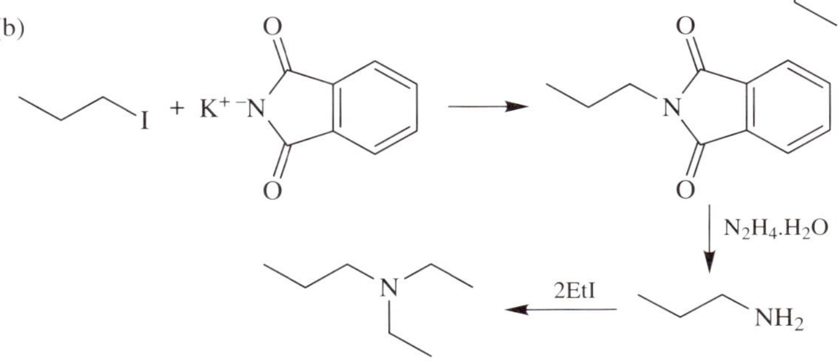

(c)

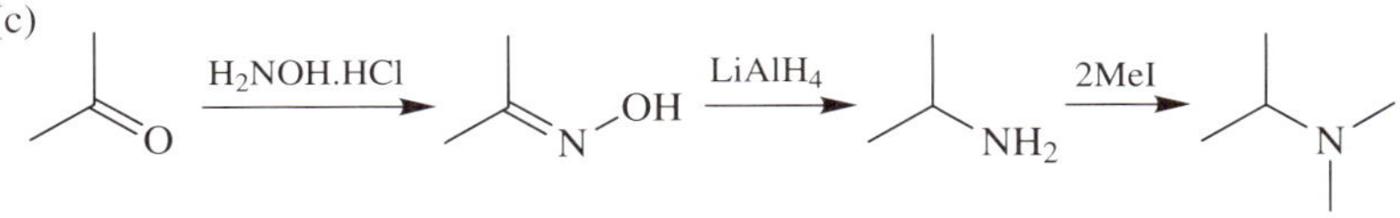

(d)

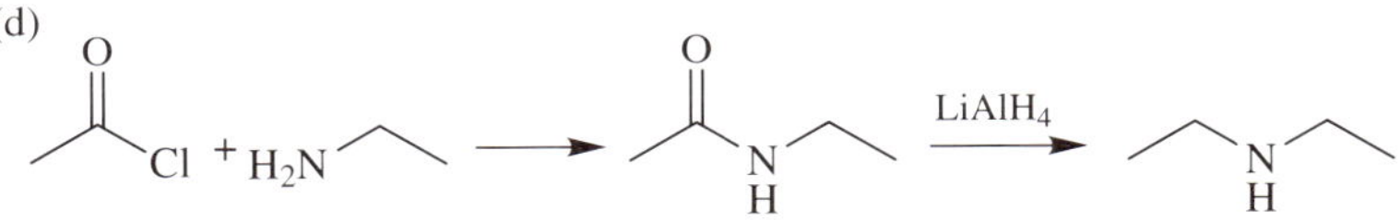

2.8. (a) *t*-Butyl bromide (Me_3CBr) readily undergoes elimination to form an alkene ($Me_2C{=}CH_2$). (b) Neopentyl bromide (Me_3CCH_2Br) undergoes a 1,2-shift with the formation of an alkene rearrangement product:

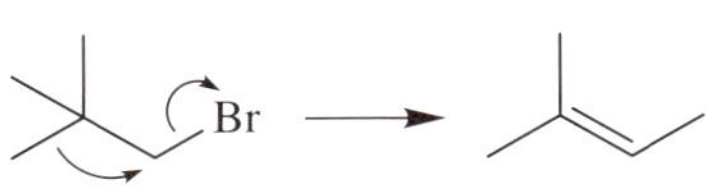

The amines are best prepared by the Hofmann degradation of the homologous amides, which may be obtained from the acyl chloride and ammonia:

(a)

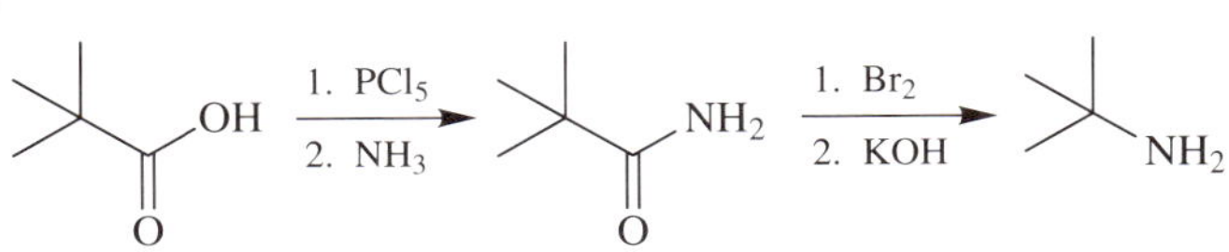

(b)

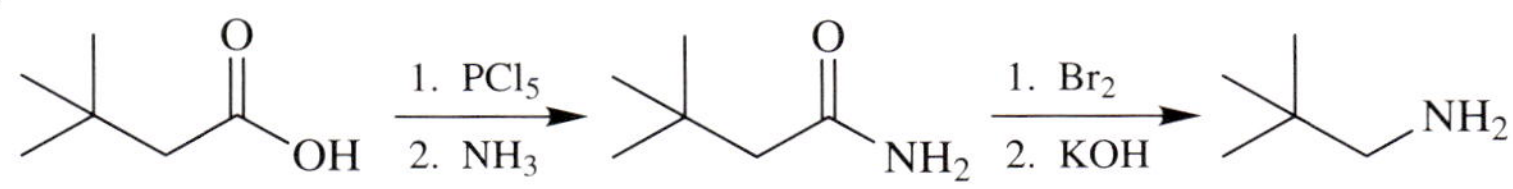

2.9. (a) A = $MeNHCO_2Et$; (b) B = EtNHCOMe; (c) C = **1**; (d) D = $H_2NCH_2CH_2CO_2H$; (e) E = $MeN(CH_2CO_2H)_2$.

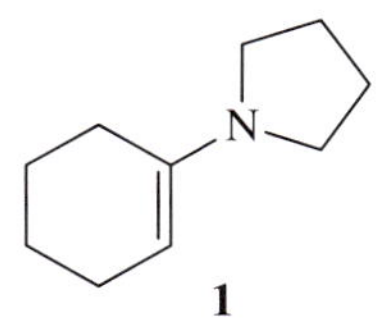

1

2.10. The pyrrolic nitrogen (a), in which the lone pair is involved in the imidazole heteroaromatic system, is non-basic. The middle nitrogen (b) is the imide nitrogen in which the lone pair is in an sp^2 hydrbridized orbital. The side-chain amine (c) is the most basic.

2.11. The amide (a) is the least basic (lone pair conjugated with the carbonyl group). The pyridine (c) nitrogen, with the lone pair in an sp^2 hybridized orbital, is less basic than the piperidine (d). The most basic is the quinuclidine (b), which is a tertiary amine in which the ring system ties the alkyl groups back, exposing the lone pair.

2.12. The reaction of the amine with nitrous acid gives a diazonium salt and thence a carbocation. The product is formed by the 1,2-shift of a methyl group:

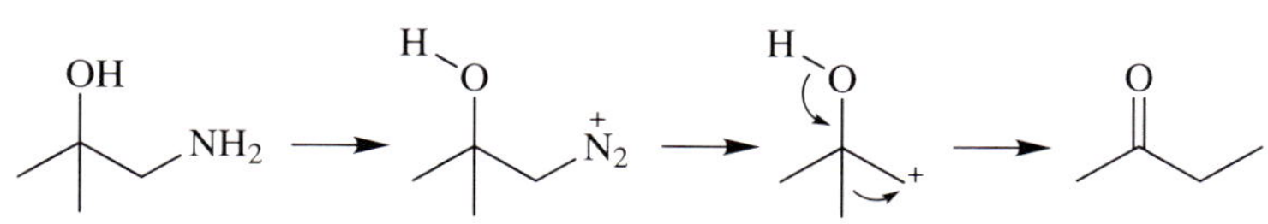

Chapter 3

3.1. The stereochemistry of the addition reactions to the alkene of α-pinene is dominated by the geminal methyl groups of the four-membered ring, which hinder addition from the same face. These methyl groups hinder the approach of a nucleophile to the rear of the epoxide. The fragmentation to form sobrerol also involves the release of the strain of the cyclobutane ring.

(a) 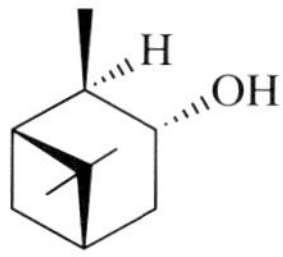

(b) 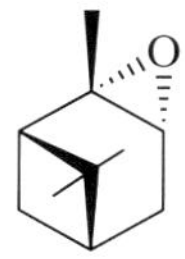

(c) 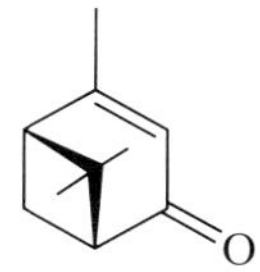

(d) 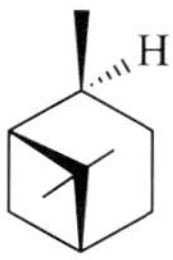

(e) 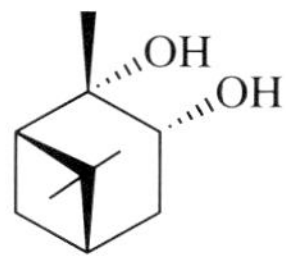

(f) 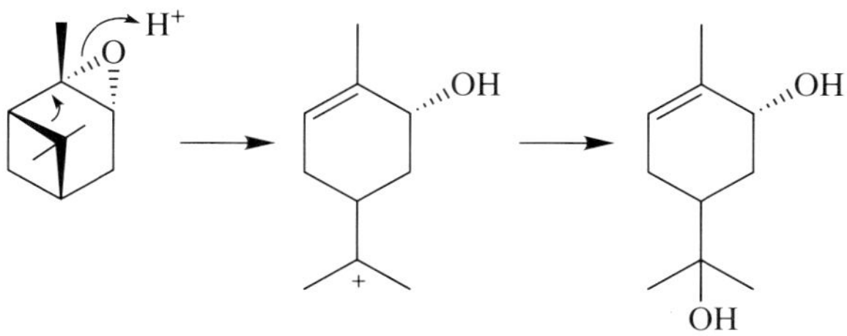

3.2.

(a) $MeC{\equiv}CH \xrightarrow[H_2SO_4]{HgSO_4} MeCOMe$

(b) $MeC{\equiv}CH \xrightarrow[2.\ H_2O_2,\ NaOH]{1.\ BH_3.THF} MeCH_2CHO$

(c) $MeC{\equiv}CH \xrightarrow[liq.\ NH_3]{NaNH_2} MeC{\equiv}C^-\ Na^+ \xrightarrow{HCHO} MeC{\equiv}CCH_2OH$

3.3.

(a) $CCl_3CHO + {}^2H_2O \longrightarrow C^2HCl_3$

(b) $CH_3COCH_3 + NaO^2H + {}^2H_2O \longrightarrow C^2H_3COC^2H_3$

(c) $CH_3CO_2Na + NaO^2H + {}^2H_2O \longrightarrow C^2H_3CO_2Na \xrightarrow{{}^2HCl} C^2H_3CO_2{}^2H$

Alternatively:

$C_3O_2 + {}^2H_2O \longrightarrow C^2H_2(CO_2{}^2H)_2 \xrightarrow{heat,\ 150\,°C} C^2H_3CO_2{}^2H$

3.4.

(a) $MeCHO + NaBH_4\ (or\ LiAlH_4) \longrightarrow MeCH_2OH$

(b) $MeCHO + CrO_3 \longrightarrow MeCO_2H \xrightarrow{PCl_5} MeCOCl$

(c) $\mathrm{MeCHO + H_2NOH \longrightarrow MeCH{=}NOH \xrightarrow{LiAlH_4} MeCH_2NH_2}$

(d) $\mathrm{MeCHO + EtMgI \longrightarrow MeCH(OH)Et \xrightarrow{CrO_3} MeCOEt}$

(e) $\mathrm{MeCHO + MeCO_2Et \xrightarrow{NaOEt} MeCH{=}CHCO_2Et \xrightarrow{NaOH} MeCH{=}CHCO_2H}$

(f) $\mathrm{MeCHO + HCN \longrightarrow MeCH(OH)CN \xrightarrow{H_3O^+} MeCH(OH)CO_2H}$

3.5.

A = 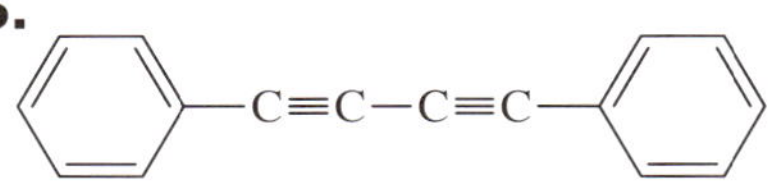

B = 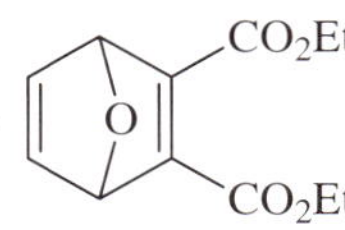

C = 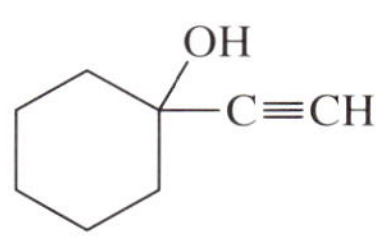

D = 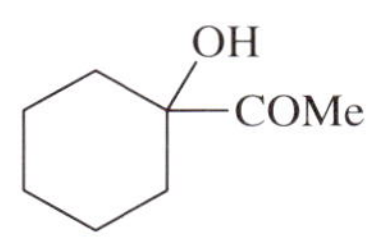

E = 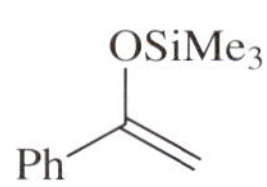

F = 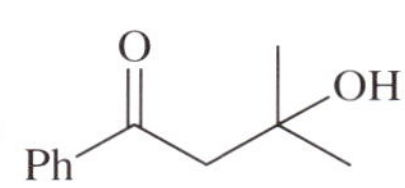

G = 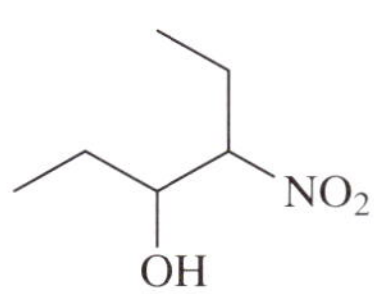

H = 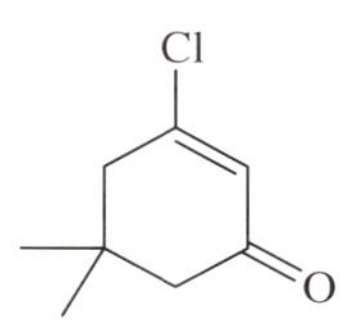

I = 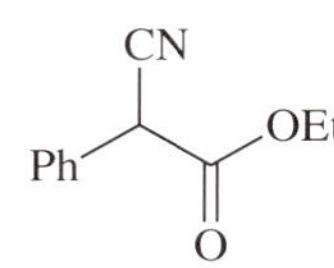

3.6. (i) Me_2CHBr; (ii) NaOEt; (iii) $CH_2{=}CHCN$; (iv) NaOEt; (v) H_3O^+; (vi) EtOH; (vii) H_3O^+; (viii) NaOEt.

3.7.

(a)

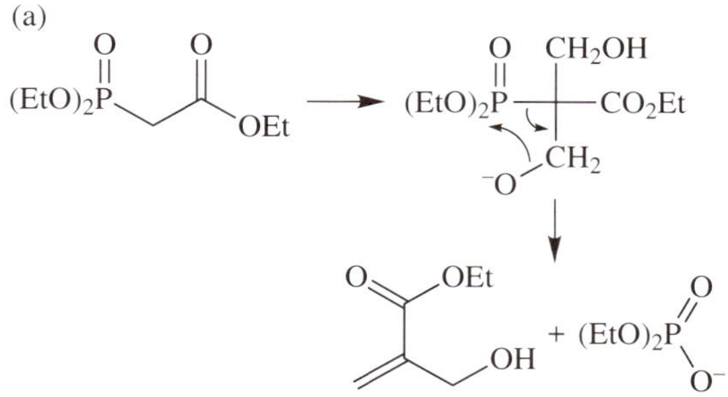

(b)

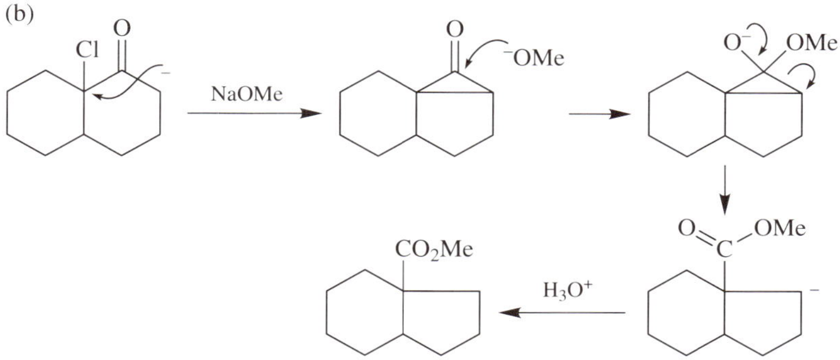

(c)

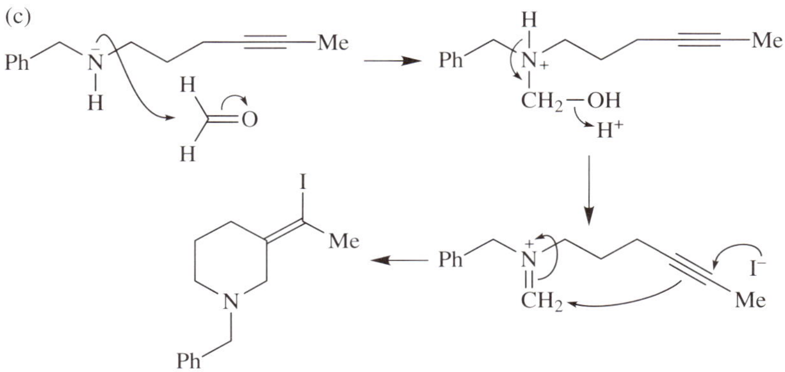

3.8.

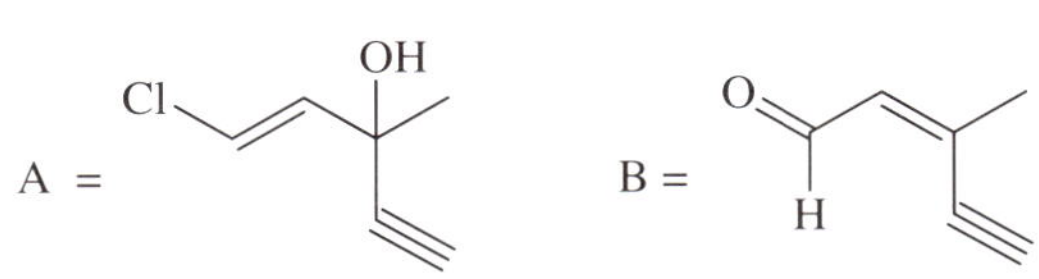

3.9.

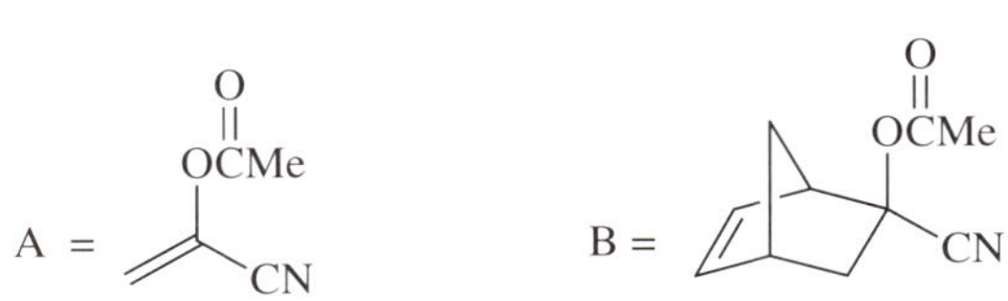

3.10.

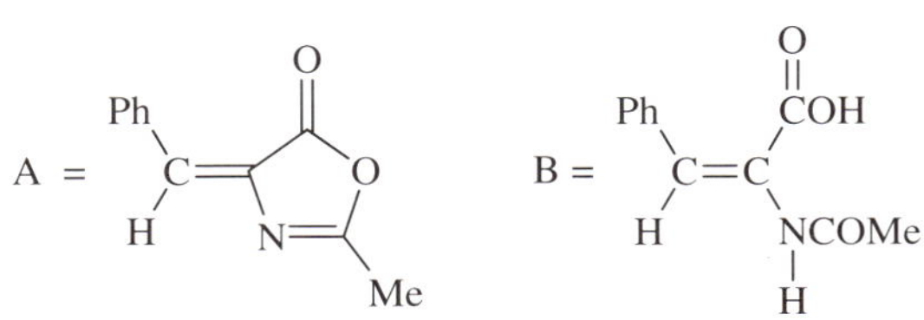

3.11.

A = 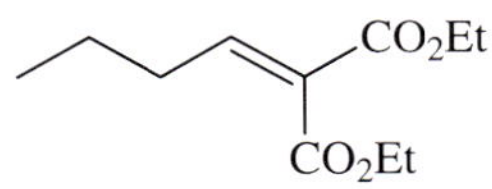 B = 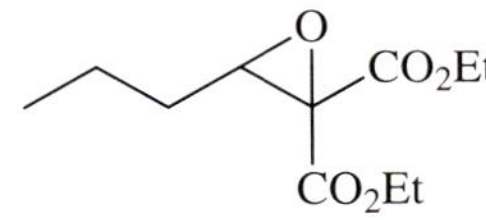

C = 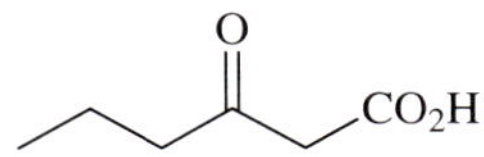

Chapter 4

4.1. Compound (d) (azulene) is a 10π cyclic conjugated system.

4.2.

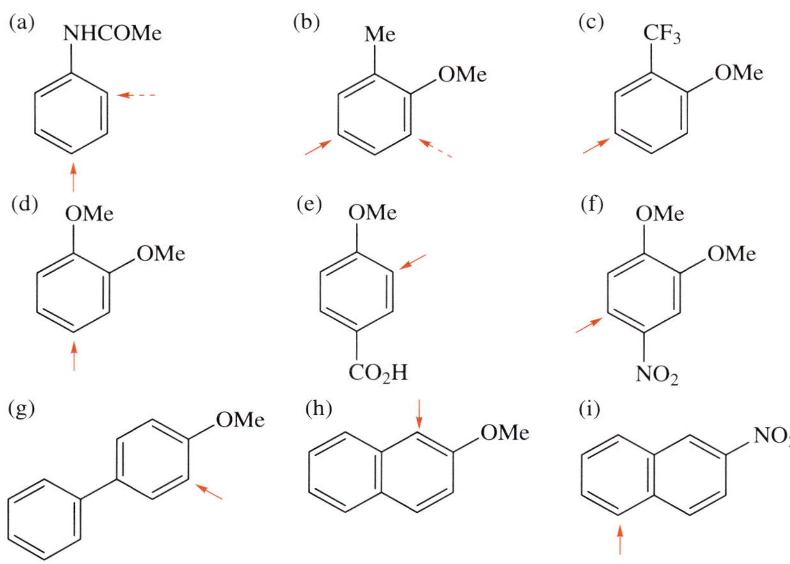

4.3.

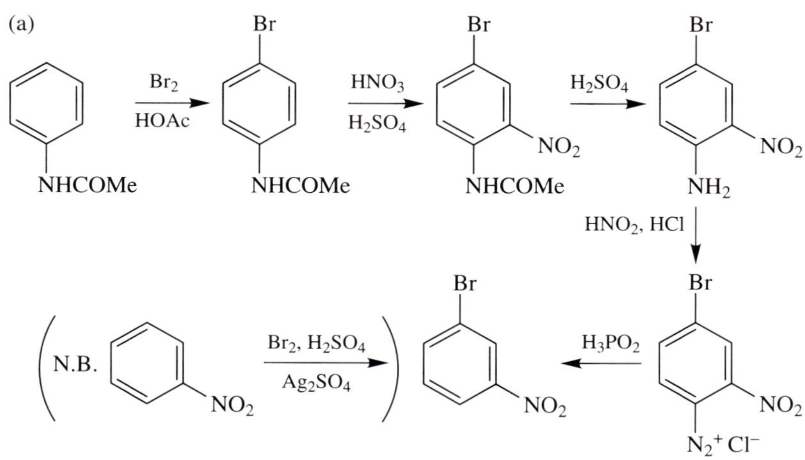

(b)

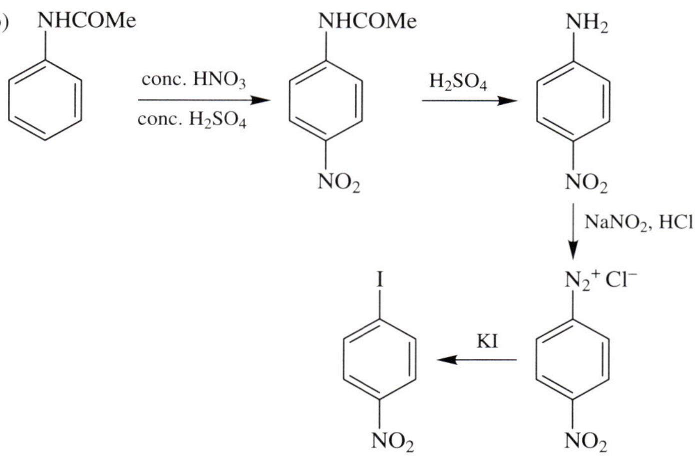

(c)

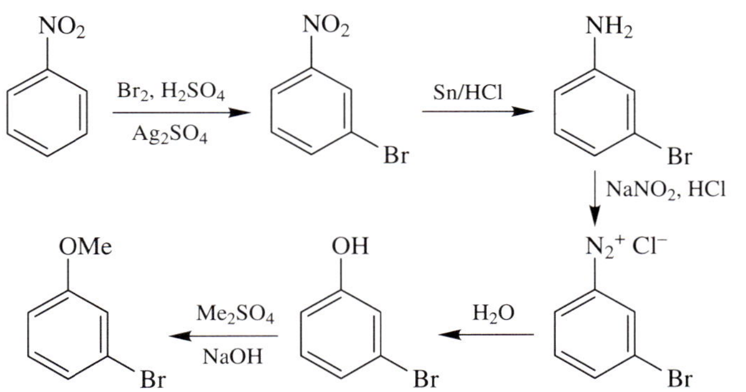

(d)

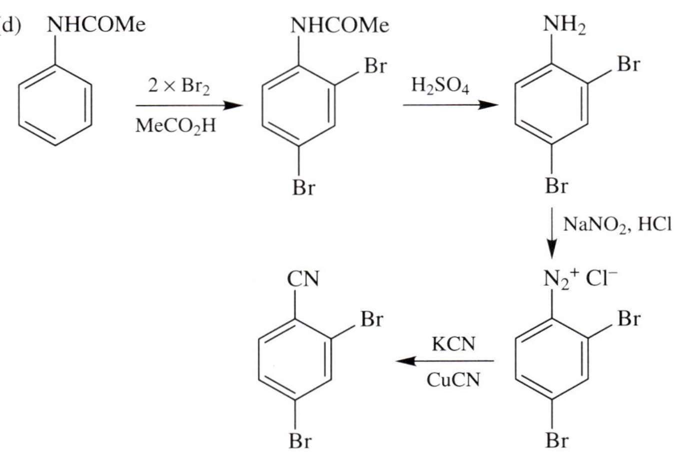

4.4.

(a)

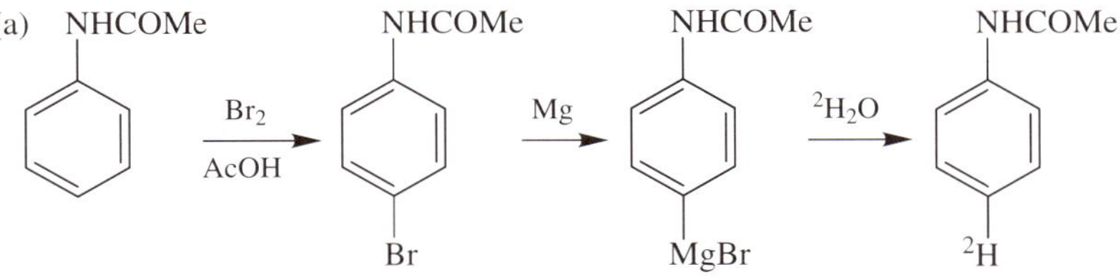

(b)

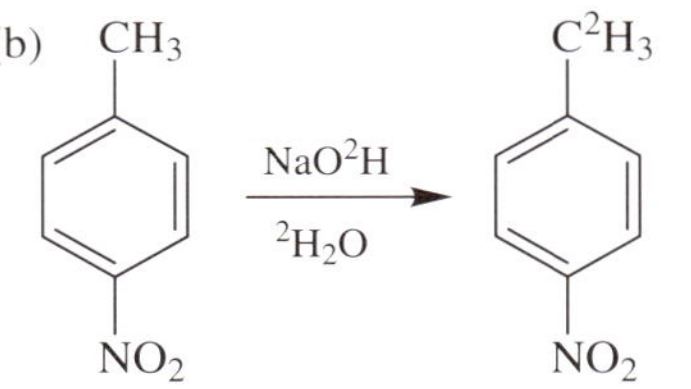

4.5. (a) 1,2-Dinitrobenzene:

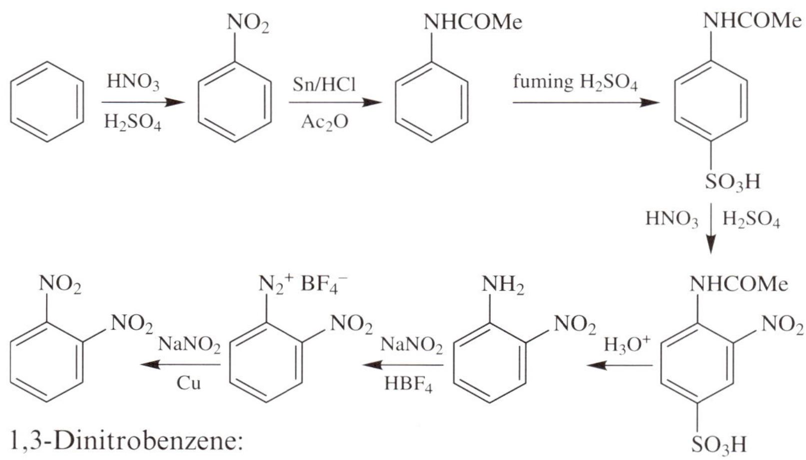

1,3-Dinitrobenzene:

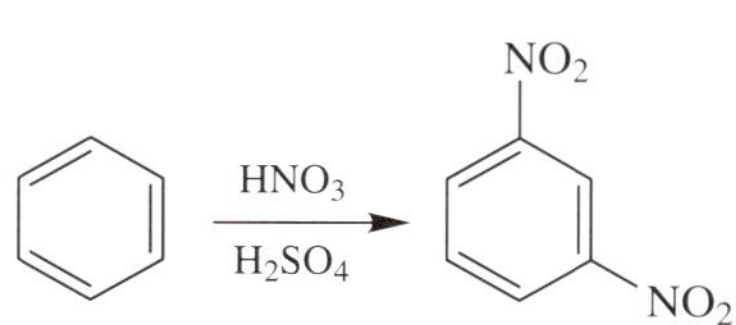

1,4-Dinitrobenzene:

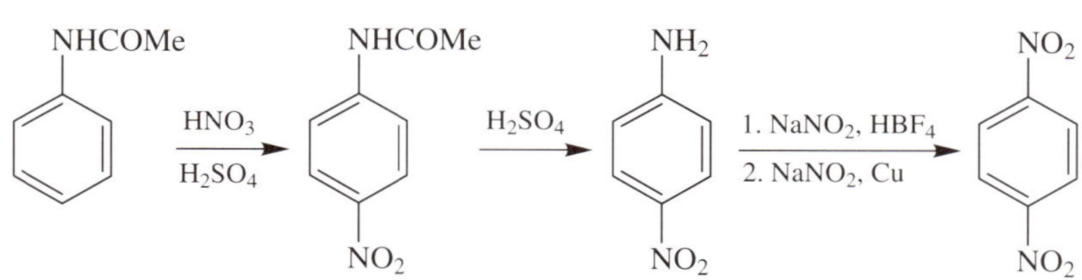

(b)

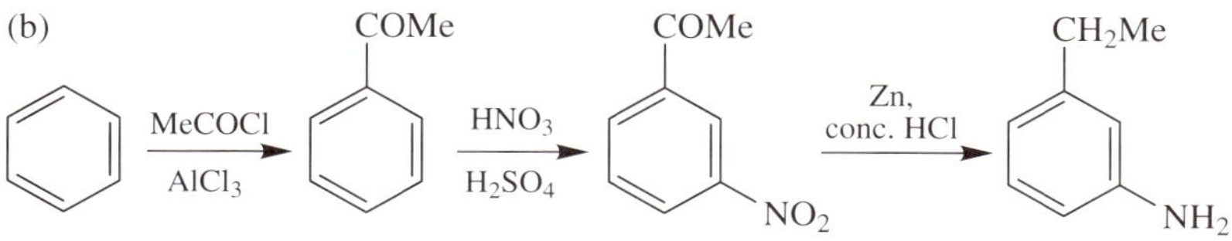

(c)

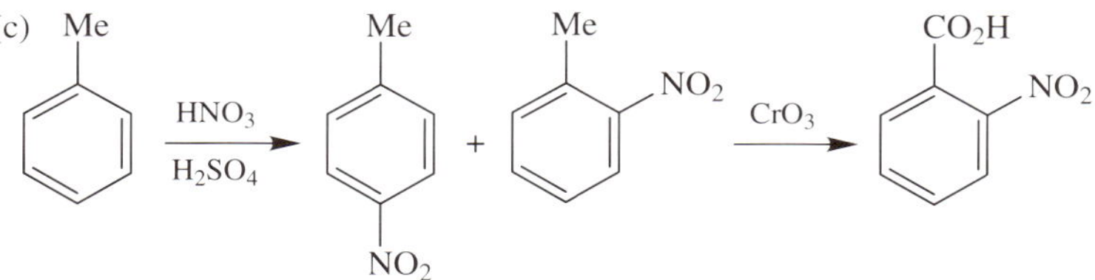

(d)

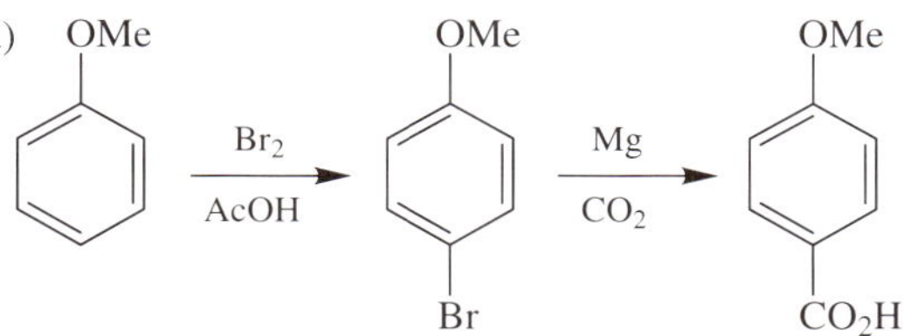

4.6. If possible, start with the activating substituent present. The *para* substituent should be introduced first and then the *ortho* substituent:

(a)

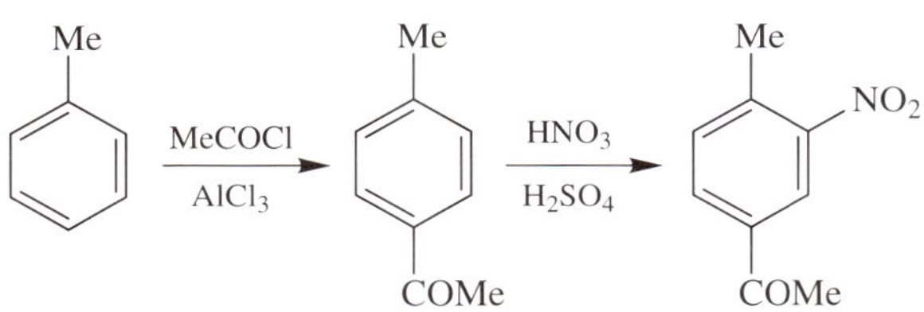

(b)

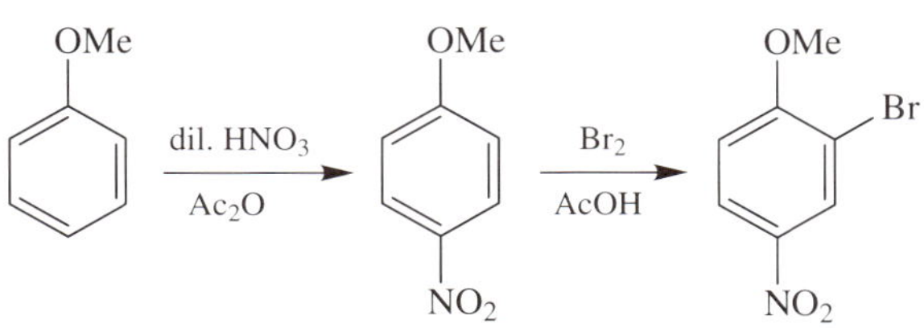

A cinnamic acid group directs a substituent to enter the *ortho* position and hence the benzaldehyde should be nitrated first and then the cinnamic acid group synthesized:

(c)

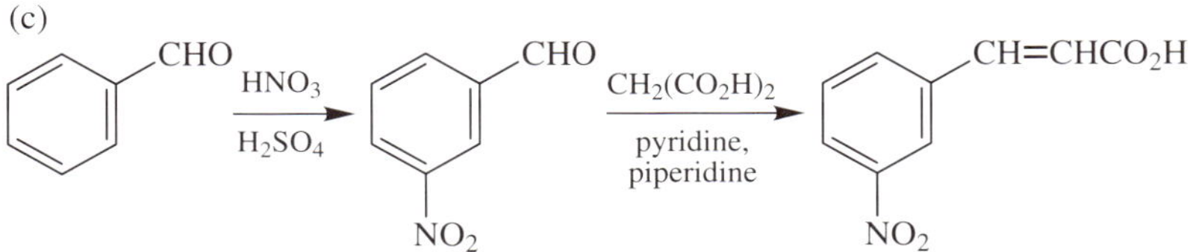

4.7.

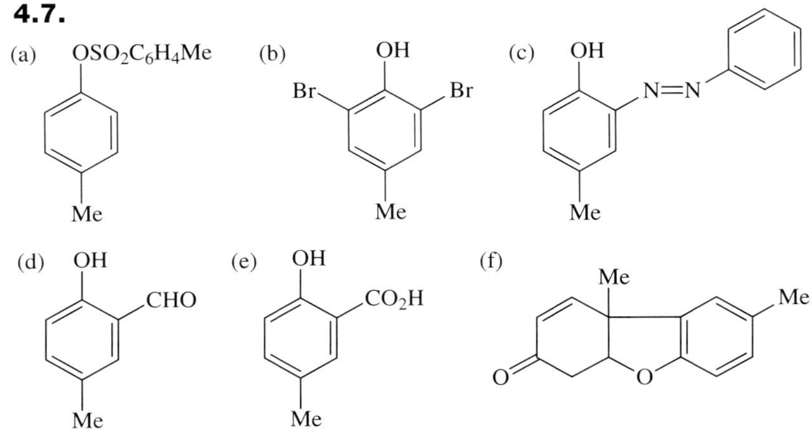

4.8. (a) *p*-Nitrophenol. (b) The lone pair on the nitrogen is conjugated with the ring in **5** whereas in **6** steric hindrance by the 2,6-dimethyl groups twists the NMe_2 out of conjugation with the ring, thus allowing the lone pair to act as a base. (c) In **7** the electron-withdrawing nitro groups make the hydroxyl proton acidic. The phenoxide anion which is formed is then stabilized by delocalization over the nitro groups. In **8** the aromaticity of the pyrimidine ring keeps the β-diketone as its enolate. Both carbonyl groups are electron withdrawing and make the enolate acidic. The anion is then delocalized over the two carbonyl groups.

4.9. (a) The amide group of acetanilide is conjugated with the aromatic ring and behaves as an activating substituent; it stabilizes the Wheland intermediate for substitution in the 4-position. In 2,6-dimethylacetanilide the amide grouping is twisted out of conjugation by steric interaction with the 2,6-dimethyl groups; it can no longer stabilize the Wheland intermediate and the orientation of substitution is determined by the methyl groups. (b) In alkaline solution, nucleophilic addition of a hydroxyl group takes place *para* to the nitro group to give the intermediate **1** which can then lose the methoxy group.

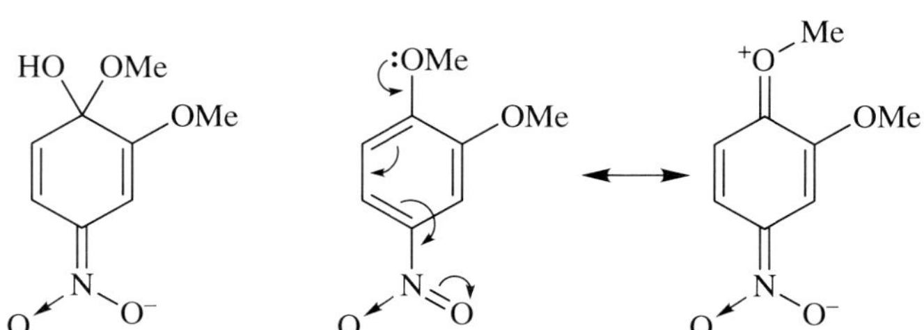

1

Under acidic conditions the more basic oxygen atom is protonated. The lone pair of the methoxy group *para* to the nitro group is delocalized over this group, whereas the *meta* methoxy group is not affected. Hence this is the more basic oxygen and is protonated and this ether is cleaved in sulfuric acid.

4.10. A lactone is a neutral oxygen-containing compound which is soluble in refluxing alkali and is regenerated by treatment with acid. Meconine contains an additional carbon atom compared to *o*-veratric acid:

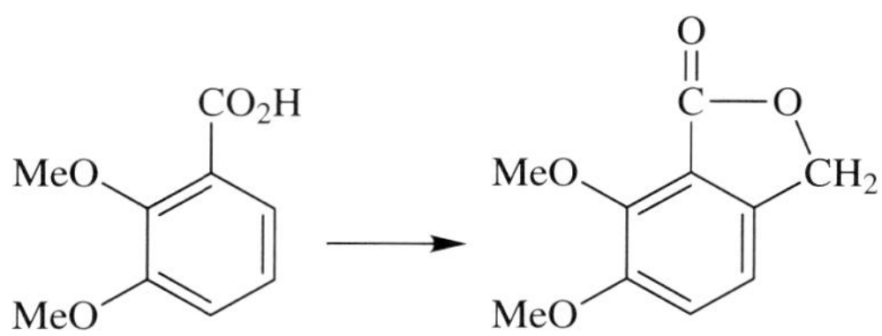

4.11. (a) (i) $(CH_2CO_2H)_2$; (ii) $AlCl_3$; (iii) Zn; (iv) conc. HCl; (v) EtOH; (vi) H_3O^+; (vii) HCO_2Et; (viii) NaOEt; (ix) H_2SO_4. (b) (i) MeCN; (ii) $ZnCl_2$; (iii) H_3O^+. (c) (i) PhCOCl; (ii) NaOH; (iii) H_3O^+.

4.12. The more nucleophilic NH_2 group reacts with the more electron-deficient carbonyl group first:

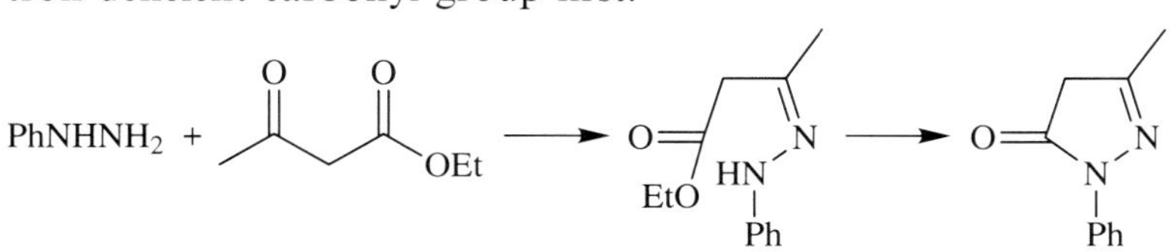

4.13.

(a) A = 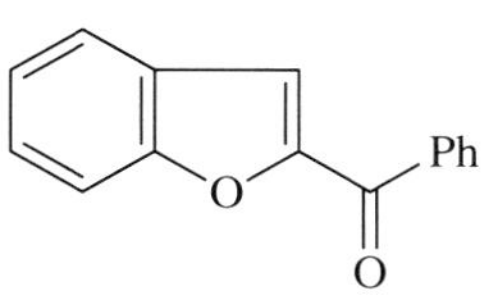

(b) B = 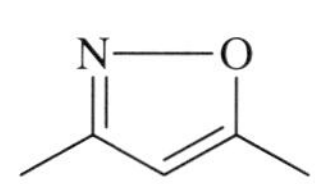

(c) C = 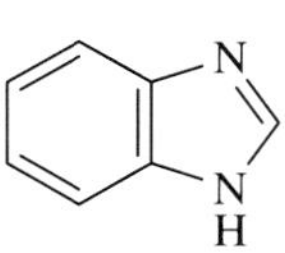

(d) D = 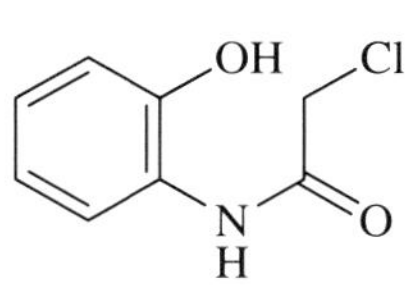

E = 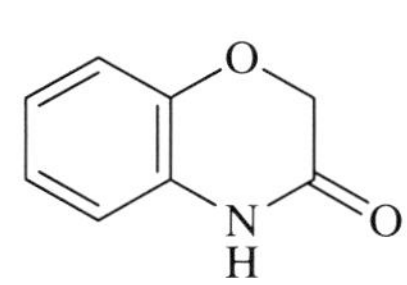

Subject Index